Design of Durable Concrete Structures

Design of Durable Concrete Structures

Editor

Rahul Barthwal

scitus academics

Design of Durable Concrete Structures

Edited by **Rahul Barthwal**

ISBN: 978-1-68117-006-0

Library of Congress Control Number: 2015931522

SCITUS Academics LLC,
www.scitusacademics.com
Box No. 4766, 616 Corporate Way,
Suite 2, Valley Cottage,
NY 10989

Printed in United States of America on Acid Free Paper ∞

Contents

Preface vii

Chapter 1 Sustainable Concrete with High Volume Ggbfs to Build Masdar City inthe Uae 1

Mohamed Elchalakana, Tarek Alyb, and Emad Abu-Aishehc

Chapter 2 Carbon Nanofiber Concrete for Damage Detection of Infrastructure 27

Y.L. Mo and Rachel Howser Roberts

Chapter 3 Effect of Alternative De-icers on the Corrosion Resistance of Reinforced Concrete Bridges and Highway Structures 51

S. O. Nwaubani and A. Katsanos

Chapter 4 Street Furniture and Amenities: Designing the User-Oriented Urban Landscape 85

Gökçen Firdevs Yücel

Chapter 5 Durable and Sustainable Road Constructions for Developing Countries 117

A.A.A. Molenaar

Chapter 6 Case Study of Remaining Service Life Assessment of a Cooling Water Intake Concrete Structure in Indonesia. 133

M. Sight Darmawan, Redo Bayuaji, N. A. Husain, and R. B. Angara

Chapter 7 Probabilistic Failure Modelling of Reinforced Concrete Structures Subjected to Chloride Penetration 175

Caio Gorla Nogueira, Edson Denner Leonel and Humberto Breves Coda

Chapter 8 A Survey of High Performance Concrete Developments in Civil Engineering Field....................213

Vatsal Patel, and Niraj Shah

Chapter 9 **Development of Optimal Structural System for Hybrid Cable-Stayed Bridges Using Ultra High Performance Concrete** **239**

Hee Seok Kim, Young Jin Kim, Won Jong Chin, Hyejin Yoon

Chapter 10 **Earthquake Damage Assessment for RC Structures Based on " Fuzzy Sets** **261**

Haoxiang He, Maolin Cong, and Yongwei Lv

Citations **291**

Index **295**

Preface

Concrete structures are designed so that they can satisfy requirements regarding safety, serviceability, durability and aesthetics throughout their design service life.

The durability aspect is a natural extension of the classical resistance verification where deterioration effects are normally neglected. The reliability is assessed through the given performance that must be delivered within the design service life, the so-called performance-based design. This approach can be adopted for a performance based on service life design. In the recent years design is related to durability through the analysis of carbonation, resistance to chloride ingress, improved freezing and thawing resistance, etc. The review of literature and some recommendations are presented referring to the design of structures aiming to attain greater durability of concrete structures. The accent is put on the theory of reliability, failure probability and service life probability. The basics of this analysis are given through the principles of performances and service life, and deterministic and scholastic methods using the lifetime safety factor.

This book gives a basic understanding of the complex set of phenomena governing durability and long term performance of concrete structures and how this forms a basis for service-life design.

Editor

Chapter 1

Sustainable Concrete with High Volume Ggbfs to Build Masdar City inthe Uae

Mohamed Elchalakan[a], Tarek Aly[b], and Emad Abu-Aisheh[c]

[a]Civil Engineering Department, Higher Colleges of Technology, Dubai Men's College, PO Box 15825, Dubai, United Arab Emirates.

[b]WorelyParsons Services, Perth, WA 6000, Australia.

[c]Consulting Services Department, Civil Engineering Unit, Saudi Aramco, Dhahran, Saudi Arabia.

ABSTRACT

Masdar City (MC) is leading the Middle East in the development of energy and resource efficient low-carbon construction in the United Arab Emirates (UAE). One of its major goals is to develop and specify materials and processes that will help reducing its environmental footprint through resource and energy conservation, as well as renewable energy generation. In 2010 MC announced on its website a prized-competition for the best proposal of "Sustainable Concrete" and "Lowest Carbon Footprint" to build MC with a total of two million cubic meter of concrete on 4 years period. This paper presents the experimental test results of 13 types of concrete mixes made with high volume of ground granulated blast furnace slag (GGBFS) cement with 50%, 60%, 70% and 80% replacement of ordinary Portland cement (OPC) to reduce the carbon emissions. A fly ash-blended mix made with 30% fly ash was also tested. The paper provides more information on the mix design parameter, full justification of CO_2 footprint, and cost reduction for each concrete type. The hardened and plastic properties and durability test parameters for each mix are presented. The results show that the slag concrete mixes significantly reduce the carbon footprint and meet the requirements of MC. An economical mix with 80% GGBFS and 20% OPC was nominated for use in the future construction of MC with 154 kg/m^3 carbon foot print.

INTRODUCTION

General

The production of 1 tonne of Portland cement requires 1.5 tonnes of raw material. The production of Portland cement is highly energy intensive, consuming 4–7 MJ of fossil fuel energy per kg (Malhotra, 1988 and Swamy, 1998), and releases approximately 1 tonne of carbon dioxide for manufacture of each tone of Portland cement. The production of cement contributes 5% of the global greenhouse gas emissions (Collins and Sanjayan, 2002). The use of slag (GGBFS), an industrial by-product which otherwise would contribute to land pollution, as a replacement for Portland cement in concrete will result in less energy for the manufacture of cement and reduce the green gas emissions due to concrete construction (Flower et al., 2005).

Slag-blended cement, a blend of ordinary Portland cement (OPC) and ground granulated blast furnace slag ("slag") has had many years use worldwide in the construction industry. In recent years, many industrial waste by-products such as slag and fly ash are rapidly becoming the main source of supplementary cementitious materials (SCM) for use in concrete manufacture. These SCMs are well known having a significant effect on reducing the concrete permeability, when properly cured, which is the main governing property for producing durable concrete (Mehta, 1984 and Hooton, 2000) suitable for the Gulf environment where sever conditions prevail. SCMs can also be used to reduce the heat generation associated with cement hydration and reduce the potential for thermal cracking in massive structural elements. The SCMs modify the microstructure of the concrete and reduce its permeability thereby reducing the penetration of water and waterborne salts into concrete thus enhancing the service life of the structure.

The inappropriate selection of cementitious materials and admixtures in mixture proportioning could have an either significant impact on cost and/or may not achieve the properties required for producing a durable concrete (i.e. high chloride resistance).

Producing sustainable concrete with a low carbon foot print was among the aims of this research. It is well known that the production of OPC produces a carbon foot print of about 1000 kg/m3 (Malhotra and Mehta, 2008). One solution to reduce the high and unaccepted construction emissions is by replacing the cement in the concrete mix (Elchalakani and Elgaali, 2010). The ground granulated blast furnace slag (GGBFS) is widely used to replace the cement to enhance the durability (Mehta, 1984 and Hooton, 2000). The GGBFS is a by-product of the steel production process (thus it is a green material), therefore, it is used here to enhance the durability and lower the carbon foot print. Thus except for the remaining small quantity of OPC used in the concrete mix, the concrete used in this project may be termed 'sustainable concrete'. To this end, this paper reports the findings of an experimental program to provide general guidelines on designing sustainable concrete mixtures suitable for use in the future construction of Masdar City in the Gulf which is a well-known harsh environment.

Within such environment, the high ambient temperature, low humidity, drying winds and dust blown salts all present great challenge to the construction of high quality concrete in the Gulf. Accordingly, special precautions need to be instituted under these extreme ambient conditions to enhance the design life and durability of MC concrete structures in service. One approach to deal with such conditions is to use high volume GGBFS concrete to increase the setting times which is beneficial for the hot climate. The GGBFS cement particles are finer (>450 m^2/kg) than the OPC ones (<350 m^2/kg). This would reduce the amount and rate of bleeding of these concretes. The reduction in bleeding together with the increase in setting times of concrete can increase the risk of plastic shrinkage cracking and may warrant special precautions during placing and finishing operations. Plastic shrinkage usually occurs within 10–12 h after placement only when concrete is exposed to unsaturated air (RH < 95%) in the presence of high speed wind and hot temperature (Collepardi, 2006 and ACI Manual, 2005).

Masdar City Requirements

Masdar City (MC) is a relatively new origination based in the UAE and is leading the Middle East in the development of energy and resource efficient low-carbon construction. One of its major goals is to develop and specify materials and processes that will help reducing its environmental footprint through resource and energy conservation, as well as renewable energy generation. In 2010 MC announced on its website a prized-competition for the best proposal of "Sustainable Concrete" and "Lowest Carbon Footprint" to build MC with the following requirements.

1. Total CO_2 emission per cubic meter of concrete should be less than Masdar baseline mix which has a carbon footprint of 192 kg/m^3.
2. Total cost of all constituent materials required per cubic meter of concrete comparable with Masdar baseline mix which has a cost of 211 AED/m^3 (1.0 USD = 3.679 AED).
3. Production capacity is anticipated at 500,000 m^3 per year for four years.
4. Concrete performance for proposed mix design, including but not limited to:

- *Workability*: slump and slump retention. A minimum slump of 150 mm is required.
- Compressive strength at 28 days not less than 40 MPa.
- *Durability*: rapid chloride penetration less than 1000 C at 28 days.

- *Hot weather*: maximum temperature of fresh concrete of 35 °C.
- *Temperature rise*: maximum concrete temperature of less than 70 °C.

This paper presents the recent research findings of 14 controlled laboratory trial mixes. It will be discussed how such mixes reduce the carbon footprint and could meet the requirements of MC. The paper provides more information on the mix design parameter, full justification of CO_2 footprint, and cost involved. The hardened and plastic properties and durability test parameters for each mix are also presented and discussed.

TEST PROGRAM

Material Properties

Table 1 shows the chemical composition of the General Portland cement type CEM I 42.5 N complying with BS EN 197-1 (2000) GP and the ground granulated blast furnace slag (GGBFS) complying to BS 6699 (1992) and fly ash to ASTM C618 Class F. The nominal target strength of concrete was 40 MPa at 28 days. Standard cubes 100 mm × 100 mm × 100 mm were prepared to BS 1881-116 (1983) and moist cured in a water tank at temperature of 25 °C. Several tests were performed to measure the durability parameters of the concrete namely rapid chloride penetration (RCP) test to ASTM C1202-97 (1997), chloride migration coefficient to NBuild 492 (1999), drying shrinkage to ASTM C157/C (2006), and water absorption (WA) test to BS 1881-122 (1983).

Table 1: Chemical composition of GP and GGBS

Constituent/property (%)									
	SiO_2	Al_2O3	Fe_2O_3	CaO	MgO	Mn_2O_3	SO_3	S	Cl
PC-CEM I 42.5 N BS EN 197 (2000)	21.29	4.89	3.42	64.16	1.41	–	2.53	–	0.010
GGBS-BS 6699 (1992)	33.22	16.12	0.72	42.42	5.53	0.30	0.32	1.30	0.009
FA-ASTM C618 Class F	70 (min)			–	–	–	5.00	N/A	–

Concrete Mix Proportions

The MC control mix had 148 kg/m^3 OPC, 222 kg/m^3 GGBFS, 137 kg/m^3 of DEWA fresh water, 720 kg/m^3 of 20 mm crushed RAK rock, 350 kg/m^3 of 10 mm crushed RAK rock, 580 kg/m^3 of 5 mm crushed RAK rock, and 300 kg dune sand. A high range water reducer addition of 4500 g/m^3 was used in the mix. The water/cement (w/c) ratio was 0.37 and the GGBFS amount represented 60% OPC replacement. The term cement refers to the binder including OPC, FA and GGBFS. Three approaches were adopted in this study to obtain better performance than that of the baseline mix. The use of high percentages of GGBFS with low water cement ratios and medium to low total cement content was the main factor used in the design mix for each group. The first approach (Group M) was to use medium cementitious material (CM) content with a total amount of 360 kg/m^3 with different w/c ratios in the range of 0.35–0.42. This approach was represented in Mixes #1, 2, 3, 4, 5, and 10. Table 2 shows the mix proportion and test results for these later mixes. The second approach (Group H) was to use a comparatively high content with a total amount in the range of 400–440 kg/m^3 with the same w/c of 0.38 and variable GGBFS content in the range of 70–80%. This approach was represented in Mixes no. 6, 7, 8, and 9. Table 3 shows the mix proportion and test results for these later mixes. The third approach (Group L) was to use very low binder content in the range of 300–340 kg/m^3 with different w/c ratios. This approach was represented in Mixes 11, 12, 13, and 14. Table 2 shows the mix proportion and workability test results for all mixes. Table 4 shows the mix proportion and test results for these later mixes.

Table 2: Group M. Trial mixes with medium cement content of 360 kg/m³

Total cement content: 360 kg							
	Trial mix	1	2	3	4	5	10
	Ref. mix	1227	1228	1229	1230	1231	1236
General details	Grade (N)	40	40	40	40	40	40
	Cement (kg)	360	360	360	360	360	360
	GGBS %	0	70%	80%	80%	80%	50%
	Fly ash %	–	–	–	–	–	30%
	w/c	0.38	0.38	0.38	0.35	0.42	0.38
	Admixture (g/m³)	8800	8200	6800	9200	6000	5200
Slump (mm)	Initial	195	235	200	220	235	210
	30 min	185	240	80	230	175	185
	60 min	150	245	45	230	75	140
Temperature (°C)	Initial	29.5	28.0	28.0	26.5	27.0	26.5
	30 min	27.5	27.0	27.0	26.0	25.5	25.5
	60 min	27.0	26.5	26.5	24.5	24.5	25.0
Average-compressive strength (N/mm²)	1 day	26.0	9.3	9.3	11.3	9.5	7.0
	3 days	57.0	35.5	32.8	36.5	28.0	28.5
	7 days	63.0	55.5	50.3	54.8	41.0	37.3
	28 days	74.3	68.0	66.0	68.3	54.0	51.3
Durability (28 days)	RCP (C)	1971	465	397	383	402	331
		1732	486	411	377	409	314
	Water absorption %	1.62	1.49	1.35	1.24	1.68	1.05
		1.67	1.45	1.46	1.09	1.71	1.16
Carbon	Carbon (kg/m³)	386	183	153	154	153	147
Cost (AED)	Cost (AED)	222	229	223	235	220	217

Table 3: Group H. Trial mixes with high volume cement content of 400–440 kg/m³

High volume cement content: 400 kg and 440 kg					
	Trial mix	6	7	8	9
	Ref. mix	1232	1233	1234	1235
General details	Grade (N)	40	40	40	40
	Cement (kg)	400	400	440	440
	GGBS %	70%	80%	70%	80%
	Fly ash %	–	–	–	–
	w/c	0.38	0.38	0.38	0.38
	Admixture (g/m^3)	5200	5600	5000	5200
Slump (mm)	Initial	215	235	230	220
	30 min	100	220	195	110
	60 min	35	115	60	40
Temperature (°C)	Initial	26.0	27.0	26.5	26.5
	30 min	25.0	25.5	25.0	25.0
	60 min	24.5	24.5	24.6	24.5
Average-compressive strength (N/mm^2)	1 day	11.8	9.0	9.5	9.0
	3 days	33.3	34.5	35.3	32.0
	7 days	49.8	43.0	43.8	37.5
	28 days	60.8	56.3	55.8	49.5
Durability (28 days)	RCP (C)	551	329	525	479
		553	364	513	491
	Water absorption %	1.58	1.67	1.65	1.90
		1.77	1.75	1.55	1.96
Carbon	Carbon (kg/m^3)	196	164	209	176
Cost (AED)	Cost (AED)	223	227	231	234

Table 4: Group L. Trial mixes with very low cement content of 300–340 kg/m^3

Low cement content: 300 kg and 340 kg (60% GGBS)						
	Trial mix #	11	13	12	14
	Ref. mix	1237	1237 B	1238	1238 B
General details	Grade (N)	40	40	40	40
	Cement (kg)	340	340	300	300
	GGBS %	60%	60%	60%	60%
	Fly ash %	–	–	–	–
	w/c	0.40	0.38	0.40	0.38
	Admixture (g/m^3)	7600	7500	9200	8590
Slump (mm)	Initial	225	230	225	215
	30 min	220	235	220	235
	60 min	210	240	220	215
Temperature (°C)	Initial	27.5	30.5	27.0	31.0
	30 min	26.5	29.5	26.0	30.0
	60 min	26.0	29.0	25.5	29.5
Average-compressive strength	1 day	12.8	–	18.0	–
	3 days	41.0	–	36.3	–
	7 days	52.5	61.8	48.0	56.8
	28 days	67.5	78.5	65.5	72.5
Durability (28 days)	RCP (C)	631	522	1421	504
		595	549	1606	481
	Water absorption %	1.40	1.36	1.17	0.96
		1.39	1.34	1.19	0.93
Carbon	Carbon (kg/m^3)	202	202	184	184
Cost	Cost (AED)	220	219	217	214

TEST RESULTS

Group M (360 Kg/M^3)

Fig. 1 and Fig. 2 show the results of compressive strength development for Group M which includes Mixes #1, 2, 3, 3, 5, and 10 where the total binder content is constant at 360 kg/m^3. As expected Mix 1 with 0% OPC replacement had the highest mechanical strength rate while Mix 10 had the lowest rate with 50% FA + 30% GGBFS (80% OPC replacement). The 28-day compressive strength was 74.3 and 51.3 MPa, for Mix 1, and 10, respectively. Mix 4 with 80% GGBFS (w/c = 0.35) had relatively good compressive strength rate and achieved 68.3 MPa strength at 28-day of concrete age. Mix 5 with 80% GGBFS (w/c = 0.42) had relatively low compressive strength rate (54 MPa strength at 28-day). It is obvious from the results that Mix 4 had good compressive strength performance due to the lowest w/c ratio in the mix.

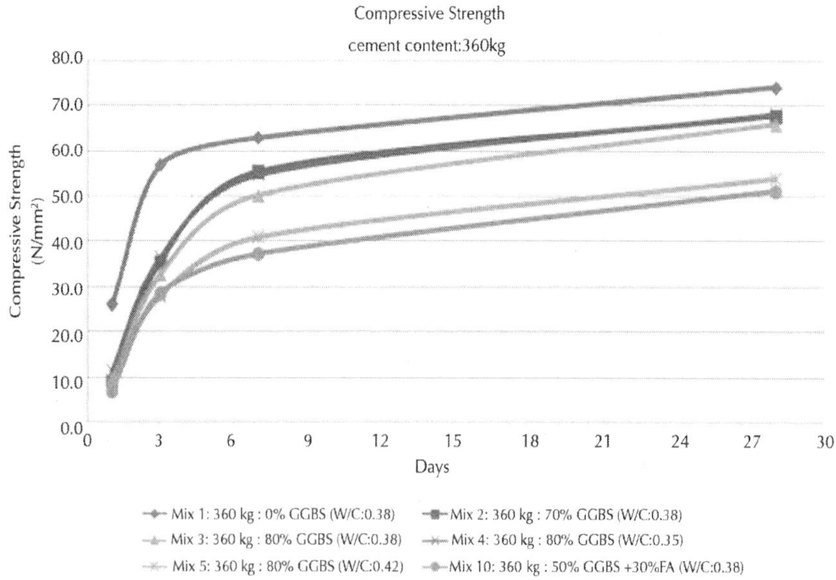

Figure 1: Mechanical strengths for Group M (360 kg/m^3).

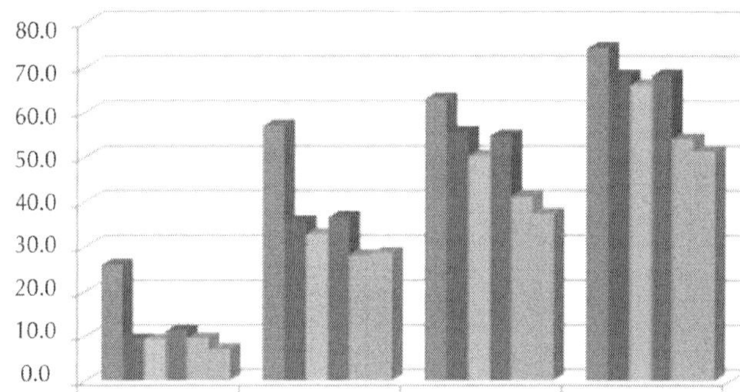

	Day 1	Days 3	Days 7	Days 28
360kg(w/c:0.38)	26.0	57.0	63.0	74.3
360kg 70%GGBS (w/c:0.38)	9.3	35.5	55.5	68.0
360kg 80%GGBS (w/c:0.38)	9.3	32.8	50.3	66.0
360kg 80%GGBS (w/c:0.35)	11.3	36.5	54.8	68.3
360kg 80%GGBS (w/c:0.42)	9.5	28.0	41.0	54.0
360kg 50%GGBS + 30%FA (w/c: 0.38)	7.0	28.5	37.3	51.3

Figure 2: Cube axial compressive strength results for Group M (360 kg/m^3).

Fig. 3 and Fig. 4 show the RCP and WA durability test results for all group of mixes of which Group M with 360 kg/m^3, respectively. It can be seen that the Mix 1 had the worst performance where it had 1971 (>1000) C and 1.67% water absorption. Mixes #2, 3, 4, 5, and 10 all can be classified as low (<1000 C) to ASTM C1202-97 (1997). Mix 10 had the best performance where it had 331 (<1000) C and 1.16% water absorption. Mix 4 had 383 (<1000) C and 1.24% water absorption. The effect of GGBFS on the resistance to chloride penetrability is clearly shown in the figure where the Mix 1 with 0% GGBFS had the worst performance. Fig. 5 shows the workability test results of the trial mixes for Group M. It is seen that Mixes 3, 5 and 10 failed to achieve the required slump of 150 mm at 60 min. The amount of the admixture for Mix 3, 5 and 10 was insufficient and of the order of 6800, 6000, 5200 g/m^3, respectively. Mixes 2 and 4 with 8200 and 9200 g/m^3 of admixture achieved higher slumps than the required.

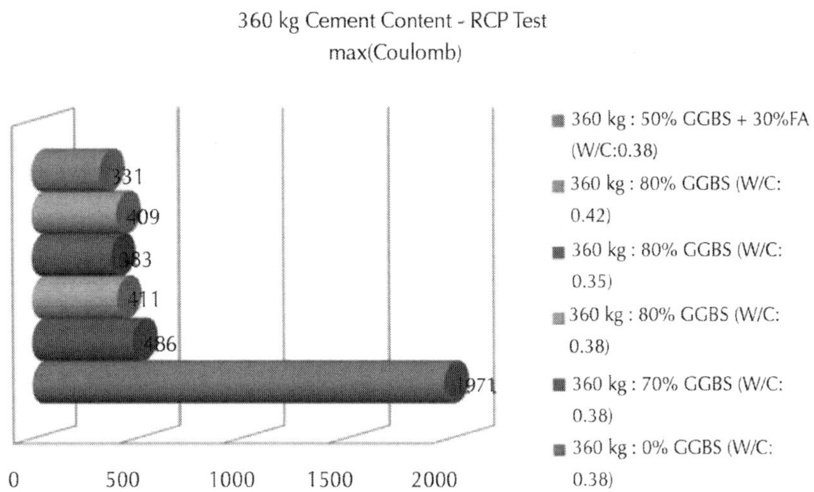

Figure 3: RCP test results for Group M (360 kg/m^3).

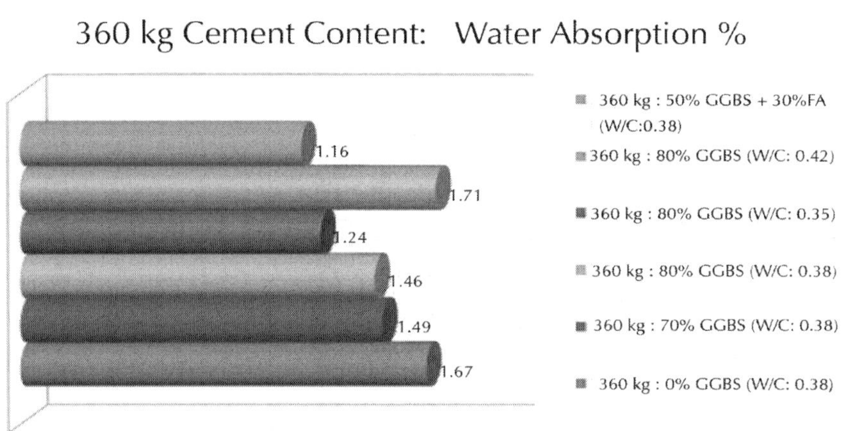

Figure 4: WA test results for Group M (360 kg/m^3).

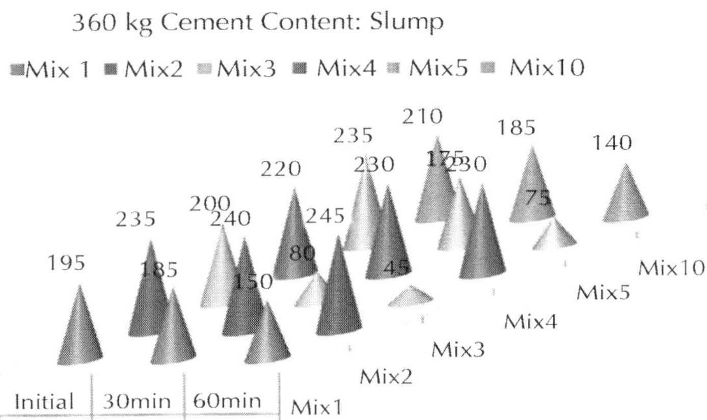

	Initial	30min	60min
Mix1	195	185	150
Mix2	235	240	245
Mix3	200	80	45
Mix4	220	230	230
Mix5	235	175	75
Mix10	210	185	140

Figure 5: Slump test results for Group M (360 kg/m³).

Group H (400–440 Kg/M³)

Fig. 6 and Fig. 7 show the compressive strength test results for Group H which includes Mixes 6, 7, 8, and 9 where the total binder content is in the range of 400–440 kg/m³. It is seen that Mix 6 with 70% GGBFS had the highest compressive strength rate while Mix 9 had the lowest rate with 80% GGBFS. The 28 days strength was 60.8 and 49.5 MPa, for Mix 6 and 9, respectively. Fig. 8 and Fig. 9 show the RCP and WA durability test results for the Group H mixes with 400–440 kg/m³, respectively. It can be seen in Fig. 8 that all the mixes in Group H achieved below 600 C. In Fig. 9, Mix 9 had experienced the highest WA performance among all the Group H mixes with 1.93%. Table 3 also shows the workability test results of the trial mixes for Group H. It is seen although all the mixes achieved over than 150 mm initial slump, they all failed to achieve the required final slump at 60 min. Thus, none of these mixes could be recommended for the construction of MC (Fig. 10).

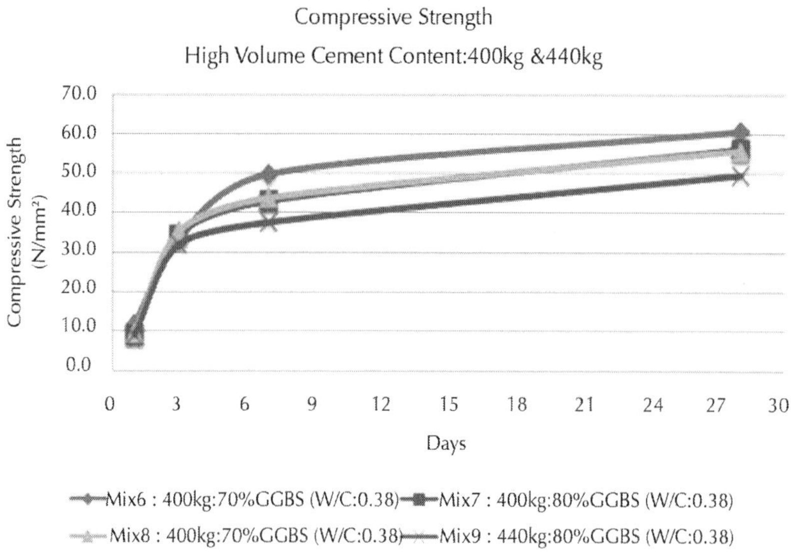

Figure 6: Mechanical strengths for Group H (400–440 kg/m³).

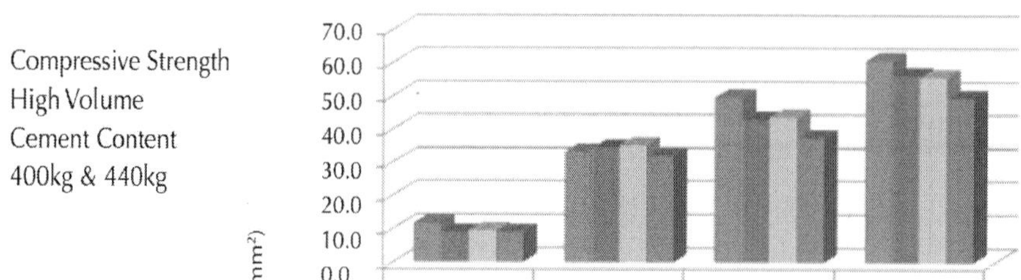

	Day 1	Days 3	Days 7	Days 28
Mix6 : 400kg : 70%GGBS (W/C: 0.38	11.8	33.3	49.8	60.8
Mix7 : 400kg : 80%GGBS (W/C: 0.38	9.0	34.5	43.0	56.3
Mix8 : 440kg : 70%GGBS (W/C: 0.38	9.5	35.3	43.8	55.8
Mix9 : 440kg : 70%GGBS (W/C: 0.38	9.0	32.0	37.5	49.5

Figure 7: Cube axial compressive strength results for Group H (400–440 kg/m³).

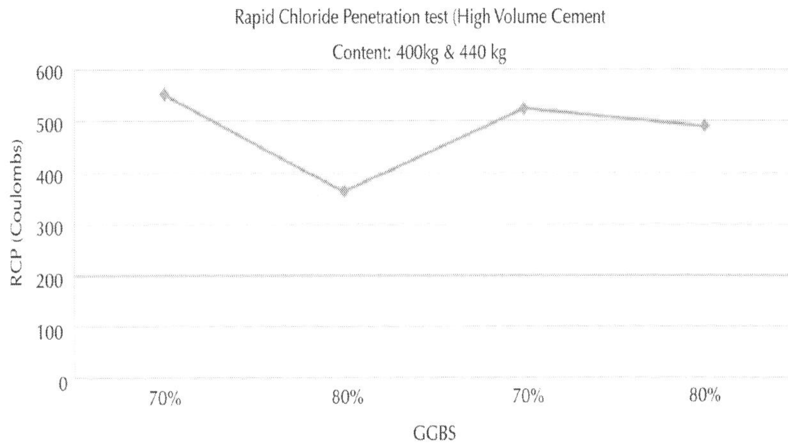

Figure 8: RCP test results for Group H (400–440 kg/m³).

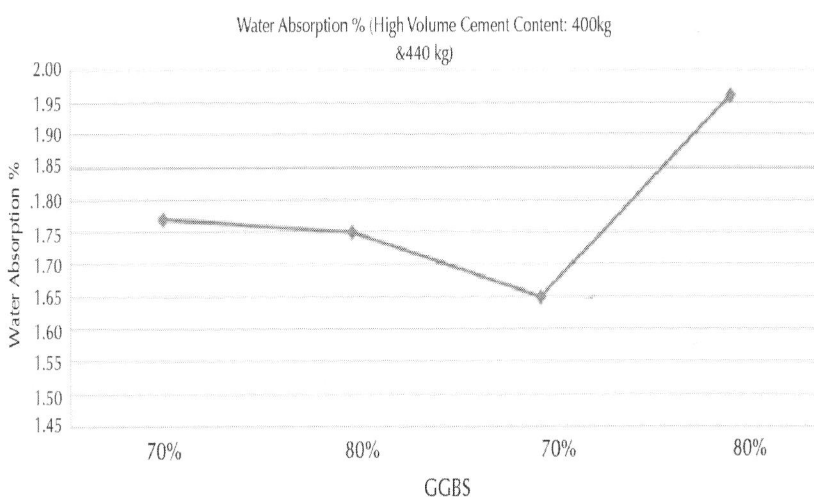

Figure 9: WA test results for Group H (400–440 kg/m³).

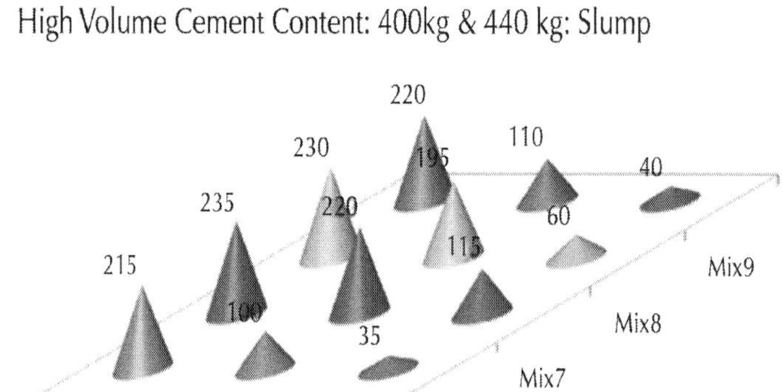

	Initial	30min	60min
Mix6	215	100	35
Mix7	235	220	115
Mix8	230	195	60
Mix9	220	110	40

Figure 10: Slump test results for Group H (400–440 kg/m³).

Group L (300–340 Kg/M³)

Fig. 11 and Fig. 12 show the compressive strength test results for Group L which includes Mix 11, 12, 13 and 14 where the total binder content is in the range of 300–340 kg/m³. It is seen that Mix 13 with 60% GGBFS and w/c = 0.38 had the highest compressive strength rate while Mix 12 had the lowest rate with 60% GGBFS and w/c = 0.4. The amount of admixture used for these mixes was 9200 and 7500 g/m³ for Mix 12 and 13 respectively. The 28-day strengths were 60.5 and 78.55 MPa, for Mix 12 and #13, respectively. Fig. 13 and Fig. 14 show the RCP and WA durability test results for Group L mixes with 300–340 kg/m³, respectively. It can be seen in Fig. 13 that all the Group L mixes achieved below 700 C, except for Mix 12 where it had poor performance and achieved relatively high value of 1513 C. Mix 14 had the best RCP and WA performance among all Group L mixes with 493 C and 0.945%, respectively. Fig. 15 shows the workability test results of the trial mixes for Group L. It is seen though all the mixes achieved the 150 mm slump at initial and final stages.

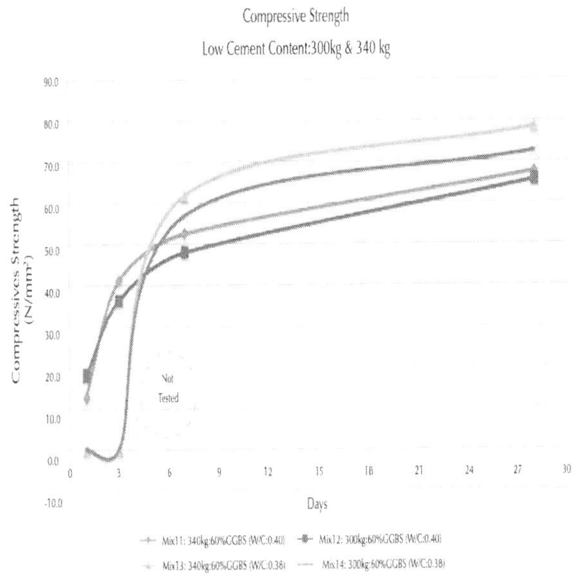

Figure 11: Mechanical strengths for Group L (300–340 kg/m³).

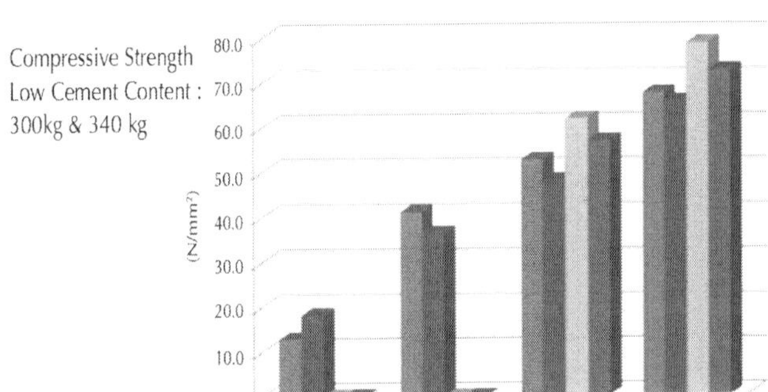

	Day 1	Days 3	Days 7	Days 28
Mix11:340kg:60%aGGBS (W/C:0.40)	12.8	41.0	52.5	67.5
Mix12:300kg:60%aGGBS (W/C:0.40)	18.0	36.3	48.0	65.5
Mix13:340kg:60%aGGBS (W/C:0.38)	0.0	0.0	61.8	78.5
Mix14:300kg:60%aGGBS (W/C:0.38)	0.0	0.0	56.8	72.5

Figure 12: Cube axial compressive strength results for Group L (300–340 kg/m³).

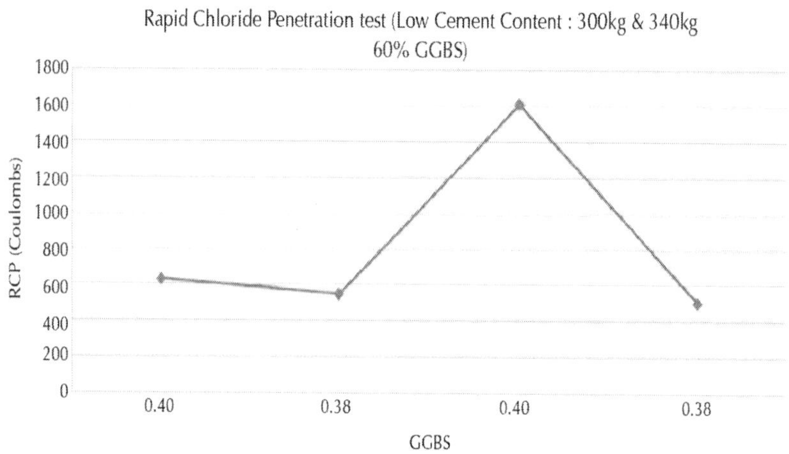

Figure 13: RCP test results for Group L (300–340 kg/m³).

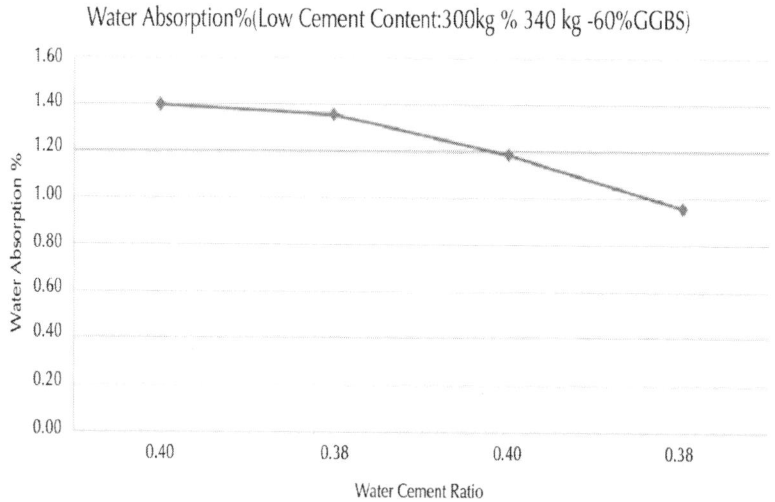

Figure 14: WA test results for Group L (300–340 kg/m³).

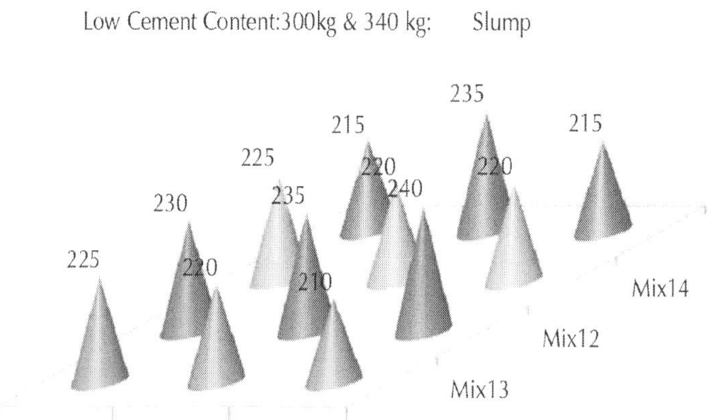

	Initial	30min	60min
Mix11	225	220	210
Mix13	230	235	240
Mix12	225	220	220
Mix14	215	235	215

Figure 15: Slump test results for Group L (300–340 kg/m^3).

SUMMARY OF RESULTS

Table 2, Table 3 and Table 4 show the initial, 30 min and 60 min setting times at temperatures below 35 °C for all the mixes. Fig. 16 shows the initial and final setting times for all the mixes except for Mix 3, 13, and 14 where their data was lost. It is seen that increasing the GGBFS from 0% (Mix 1) to 70% (Mix 2) the initial and final setting times increased by 10% and 29%, respectively. Also, Mix 12 (with low cement 300 kg, large admixture amount and 60% GGBFS) had the highest initial and final setting times of 15.3 and 21.5 h respectively. Mixes 9 (with high cement content of 440 kg, low admixture 5200 g/m^3 and 80% GGBFS) had the lowest initial and final setting times of 5.3 and 8.3 h. Fig. 16 shows that the initial and final setting times for Mix 4 are 11.4 and 16 h.

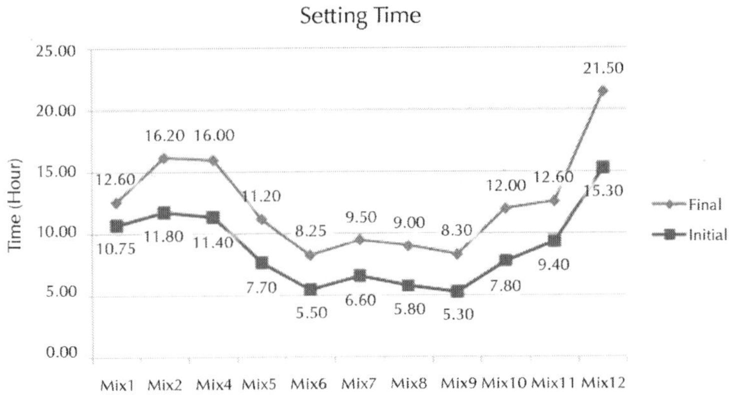

Figure 16: Setting time (initial and final) test results for all except for Mixes #3, 13, 14 (data not available).

Fig. 17 and Fig. 18 summarize the RCP and WA results for all the mixes, respectively. It is seen that the worst RCP performance was for Mix 1 where 0% GGBFS was used. Mix 9 had the worst WA performance with 1.93%, whereas Mix 10 (50% GGBFS and 30% FA) had the best RCP performance with just 322 C and 1.11% WA. Mix 14 had a good RCP performance with 493 C and the best water absorption performance of 0.95%. Fig. 19 summarizes the chloride migration coefficient (CMC) test results for all the mixes. It is seen that the trends as similar to the RCP ones where Mix 4 achieved the lowest chloride ion migration resistance where CMC = 1.05×10^{-12} m^2/s. Fig. 20 shows the 56-day drying shrinkage test results for all the mixes. Mix 10 (50% GGBFS and 30% FA) had the lowest shrinkage with just 110 micro strain. It is also seen that the trends is consistent with that of WA ones where Mix 4 and 14 achieved very low drying shrinkage at 160 and 140 micro strain. This attributed to the reduced water absorption and less paste volume which is the main source of drying shrinkage. Lower drying shrinkages will reduce the risk of cracking of concrete at the early age (Aly and Sanjayan, 2006 and Aly and Sanjayan, 2007).

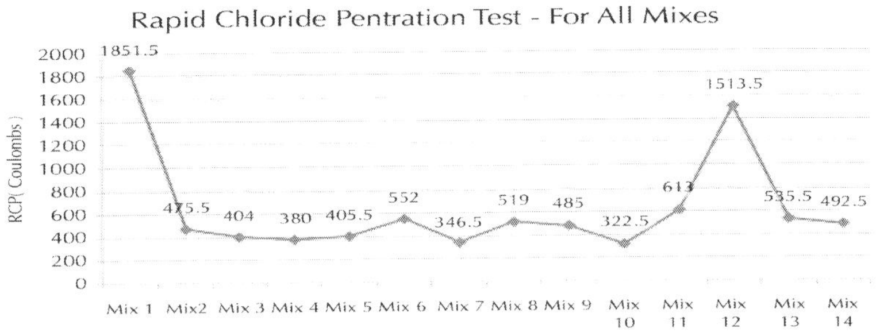

Figure 17: RCP test results for all mixes.

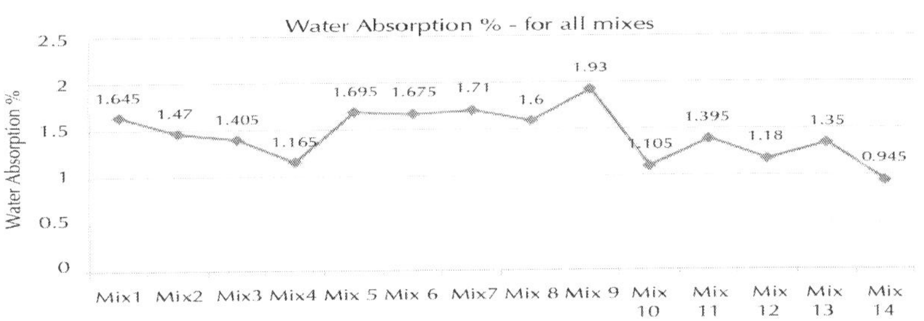

Figure 18: WA test results for all mixes.

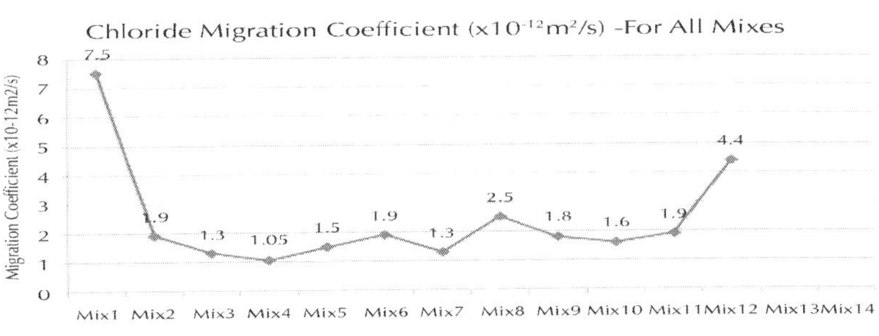

Figure 19: Chloride migration coefficient test results for all mixes.

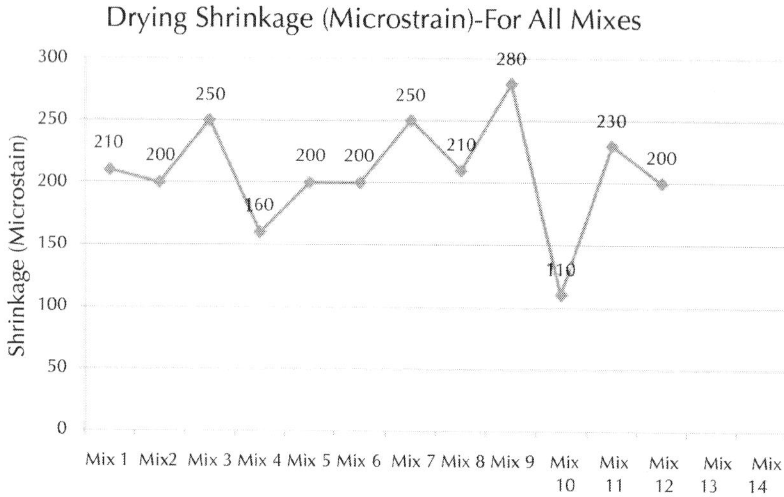

Figure 20: Drying shrinkage test results for all mixes.

Fig. 21 shows that the estimated CO_2 emission due to the production of Mix 1 with 100% OPC is 386 kg/m³ whereas the emissions for Mix 4 and 14 are 154 and 184 kg/m³, respectively.

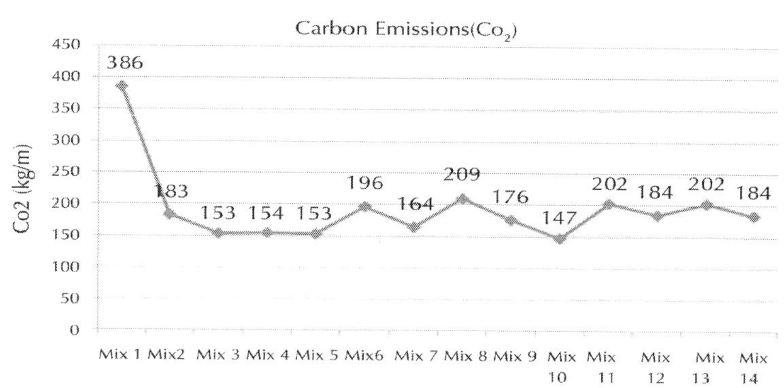

Figure 21: CO_2 emissions for all mixes.

Fig. 22 shows the values of the CO_2 intensity indicator ($c_i = c/p$) that is proposed by Damineli et al. (2010) for all the 14 mixes. Where c is the total CO_2 (kg m^{-3}) emitted to produce the concrete raw material, and p is the average compressive strength at 28 days (MPa). It is seen that all the mixes achieved a value less than the international future benchmark of $c_i = 5.0$ kg m^{-3} MP^{-1}, except for the control mix (Mix #1). Fig. 23 shows the CO_2 intensity indicator plotted against the binder intensity (b_i = b /p). Where b is the total consumption of the binder material including OPC, fly ash and GGBS (kg m$^{-3)}$. Except for the control mix, the binder intensity was in the middle range of b_i = 4.0–9.0 kg m^{-3} MP^{-1} to produce a sustainable concrete with a favorably low CO_2 intensity below 5.0 kg m^{-3} MP^{-1}.

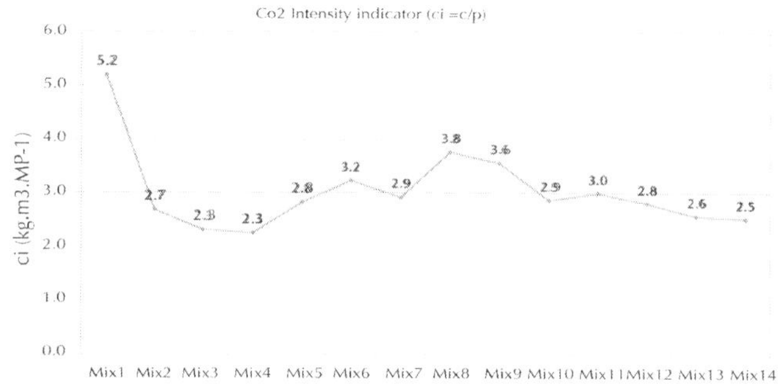

Figure 22: CO_2 intensity indicator for all mixes (c_i = c/p).

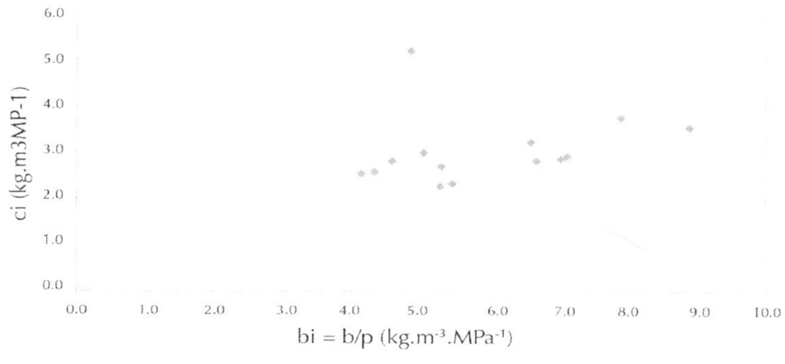

Figure 23: CO_2 intensity indicator (ci = c/p) versus binder intensity (b_i = b /p) for all mixes.

Fig. 24 shows that the estimated cost due to the production of one cubic meter of Mix 1 with 0% OPC replacement is 222 kg/m^3 whereas the cost for Mix 4 and 14 is 235 and 214 AED, respectively. Although the costs of these mixes are slightly more than MC baseline mix (<12%) they are still economical.

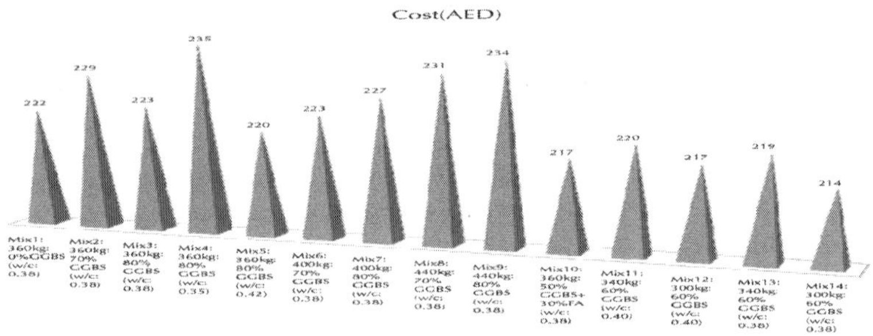

Figure 24: Cost (in AED) for all mixes.

CONCLUSIONS

Based on the plastic and hardened test results on of 14 types of concrete mixes, Mix 4 from Group M (360 kg total binder with, 80% GGBFS, 0.35 w/c) and Mix 14 from Group L (300 kg, 60% GGBFS, 0.38 w/c) are proposed for use in the construction of Masdar City in the UAE. These mixes meet the strength, durability, setting times, and workability requirements of Masdar City as well as they are cost effective compared to MC baseline mix. Although Mix 10 with 50% GGBFS and 30% FA achieved good durability and shrinkage performance as well as low carbon emissions it could not meet the workability requirements of Masdar City.

ACKNOWLEDGMENTS

The authors would like to thank Mr. Abdulazis Al Ali who is a former student of Dubai Mens College for supervising the experiments. Also thanks are given to Ready Mix Abu Dhabi's engineers and technicians for performing the trial mixes and BASF for providing the necessary admixtures.

REFERENCES

1. ACI Manual of Concrete Practice. ACI Committee 305R – Hot Weather Concreting. 2005.
2. ASTM C 1202. Standard test method for electrical indication of concrete's ability to resist chloride ion penetration. 1997.
3. ASTM C157/C. Standard test method for hardened hydraulic cement mortar and concrete. 2006.
4. BS 1881: Part 116. Method of determination of compressive strength of concrete cubes. British Standards; 1983.
5. Aly T, Sanjayan JG. Cracking tendency of concretes made with slag blended cements subjected to restrained shrinkage conditions. In: Proceedings of the 19th
6. Australasian conference on the mechanics of structures and materials. 29 November–1 December 2006, Christchurch, New Zealand; 2006; 557–62.
7. Aly T, Sanjayan JG. Factors contributing to early age shrinkage cracking of slag concretes subjected to 7-days moist curing. Materials and Structures Journal2007; 41(May (4)):633–42.
8. BS 1881: Part 118. Method of determination of flexure strength of concrete. British Standards; 1983.
9. Collins F, Sanjayan JG. The challenge of the cement industry towards the reduction of greenhouse emissions. In: Proceedings of the International Association of Bridge and Structural Engineers (IABSE) conference. September, Melbourne; 2002.
10. Collepardi M. The new concrete. Italy: Grafiche Tintoretto Publishers; 2006.
11. Damineli B, Kemeid FM, Aguiar PS, john VM. Measuring the co-efficiency of cement use. Cement and Concrete Composites 2010; 32(8):555–62.
12. Elchalakani M, Elgaali E. Strength and durability of recycled concrete from demolition and construction Wastes. In: International conference on concrete sustainability. 13–14 December, Dubai, UAE; Grey Matters; 2010. Invited Paper.
13. Hooton R. Canadian use of ground granulated blast-furnace slag as a supplementary cementing material for enhanced

performance of concrete. Canadian Journal of Civil Engineering 2000; 27(4):754–60.

14. Flower DJ, Sanjayan JG, Baweja D. Environmental impacts of concrete production and construction. In: Proceedings of the 22nd biennial conference of the concrete Institute of Australia. Melbourne; 2005; 10.
15. Malhotra VM, Mehta PK. High performance high-volume fly ash concrete for building sustainable and durable structures. 3rd ed. Ottawa, Canada: Taylor & Francis Group; 2008.
16. Malhotra VM. Use of fly ash, slag and condensed silica fume in North America and Europe. In: Ryan WG, editor. Proc concrete workshop 88. Int workshop on the use of fly ash, slag, silica fume and other siliceous materials in concreteJuly 4–6, Sydney, Australia; 1988;23–55.
17. Mehta PK. Mineral admixtures. In: Ramachandran VS, editor. Concrete admixtures handbook. Park Ridge, NJ: Noyes Publication; 1984. p. 303–33.
18. NBuild 492. Standard test method for chloride migration coefficient from non-steady state migration experiments. Espoo, Finland: NORDTEST Publishers; 1999.
19. Swamy RN. Designing concrete and concrete structures for sustainable development.In: Malhotra VM, editor. Proc sixth international conference on fly ash, slag, silica fume and other natural pozzolans in concreteBangkok, Thailand; ACI SP-178 1998;1:245–55.

Chapter 2

Carbon Nanofiber Concrete for Damage Detection of Infrastructure

Y.L. Mo[1] and Rachel Howser Roberts[1]

[1] University of Houston, USA

INTRODUCTION

Fiber research in concrete construction is an ongoing field and the use of carbon nanofibers (CNF) is examined. Fibers improve brittle materials such as concrete by enhancing tensile strength, ductility, toughness, and conductivity. Short-fiber composites are a class of strain sensor based on the concept of short electrically conducting fiber pull-out that accompanies slight and reversible crack opening. For a fiber composite to have strain sensing ability, the fibers must be more conducting than the matrix in which they are embedded, of diameter smaller than the crack length, and well dispersed. Their orientations can be random, and they do not have to touch one another. The electrical conductivity of the fibers enables the direct current (DC) electrical resistivity of the composites to change in response to strain change or temperature, allowing sensing.

NANOTECHNOLOGY IN CONCRETE

Despite the fact that nanotechnology is a relatively recent development in scientific research, the introduction of the concept is credited to Nobel Prize winner Richard Feynman from his 1959 lecture, "There's Plenty of Room at the Bottom" [1]. Feynman considered the possibility of direct manipulation of individual atoms as a powerful form of synthetic chemistry. Decades later, Feynman's concept morphed into the field of nanotechnology. According to the National Science Foundation and National Nanotechnology Initiative, the definition of nanotechnology includes three elements [2]:

- The size range of the material structures under consideration should be approximately 100 nanometers;
- The nanotechnology should have the ability to measure or transform at the Nano scale;
- There should be properties that are specific to the Nano scale as compared to the macro or micro scale.

Following this definition, in the past 25 years nanotechnology has expanded from Feynman's idea and now finds applications in fields ranging from medical devices to Nano-reinforced concrete [3, 4].

To date, the awareness and application of nanotechnology in the construction industry are increasing; however, progress is uneven in the current early stages of its practical exploitation. Bartos [5] presents three reasons for this phenomenon:

- The nature of the construction industry differs greatly from the other industries doing research in nanotechnology. The final products coming from the construction industry are not mass-produced and require relatively long service lives, differentiating it from the products from the microelectronics, information technology, and automotive industries.

- Historically, there is a very low level of investment in construction research and development.
- Research in Nano-related research and development requires very high initial capital investment

Despite these difficulties, there have been significant advances in Nano science of cementations materials with an increase in the understanding of basic phenomena in cement at the Nano scale. These include structure and mechanical properties of the hydrate phases, origins of cement cohesion, cement hydration, interfaces in concrete, and mechanisms of degradation [6]. A major nanotechnology application is to include Nano-sized reinforcement in cement-based materials such as carbon nanotubes or Nano fibers.

FIBER REINFORCED CONCRETE

Concrete, composed of fine and coarse aggregates held together by a hydrated cement binder, is one of the most important construction materials and is used in diverse project areas including house foundations, high rise tower components, highways, and dams. Hydrated cement is a brittle material that is an order of magnitude stronger in compression than in tension. To compensate for this weakness reinforcement consisting typically of rebar or fibers are added to the concrete.

The use of fibers to reinforce brittle materials can be traced back to ancient times when straw and hair was added to mud bricks. The modern development of the use of fibers in construction began in the 1960s with the addition of steel fibers to reinforced concrete structures. This was closely followed by the addition of polymeric fibers, glass fibers, and carbon fibers in the 1970s, 80s, and 90s, respectively [7].

Fibers improve brittle materials such as concrete by enhancing tensile strength, ductility, toughness, and conductivity [8-13]. Fibers are typically used in two forms: short randomly dispersed fibers in a cementations matrix or a continuous mesh of fibers used in thin sheets. Here we will focus on randomly dispersed fibers used to arrest cracks. The cracking process within concrete begins with the onset of isolated Nano cracks. These Nano cracks grow together to form localized micro cracks, which in turn grow together to form macro cracks. These macro cracks widen to form cracks visible with the naked eye. Fibers arrest these cracks by forming bridges across them. With increasing tensile stress, a bond failure eventually occurs, and the fiber will pull out of the concrete allowing the crack to widen. Fig. 1 shows the bridging action of fibers across micro and macro cracks in concrete.

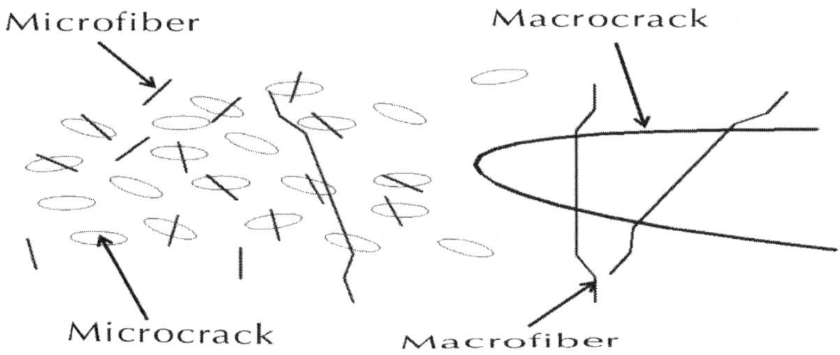

Figure 1: Bridging action of fibers across micro and macro cracks.

NANOREINFORCEMENT IN CEMENT-BASED MATERIALS.

Since the discovery of carbon nanotubes (CNT) in 1991 [14], researchers have desired to implement the unique mechanical, thermal, and electronic properties of CNT and CNF in cement-based composites. Single-wall CNT (SWCNT), multi-wall CNT (MWCNT) and CNF are grapheme ring-based materials with aspect ratios greater than 1000 and high surface areas [6, 15]. CNT and CNF have moduli of elasticity in the range of terrapascals and tensile strength on the order of gigapascal [6, 16, 17]. SWCNT consist of a single grapheme sheet wrapped into a seamless cylinder, while, as the name suggests, MWCNT inhere of multiple concentric sheets of graphene wrapped around a hollow core. CNF are cylindrical nanostructures with graphene layers arranged as stacked cones, cups, or plates. CNF are adventagious because their stacked structure presents exposed edge planes not present in CNT that intoduce increased surface area and better bond characteristics. Because of their structure, CNF are easier to produce and cost 100 times less than SWCNT [18]. Because of the increased bond surface and lower cost, CNF are more attactive than CNT for application in cement-based composites.

CNT AND CNF DISPERSION

The majority of nanoreinforced composite research has been completed on polymers containing CNT or CNF [6, 19, 20]. One of the main reasons for this is because uniform dispersion is difficult in cement-based materials. Well dispersed CNF results in uniform calcium-silicate-hydrate (CSH) gel formation, which improves the structural and electrical properties of the concrete [21]. CNT and CNF are inherently hydrophobic and are attracted due to Van der Waals forces, causing the fibers to tend to agglomerate hindering their dispersion in solvents [17, 22-24].

Several solutions have been proposed to solve this issue including dispersing the fibers through milling, ultra sonication, high shear flow, elongation flow, functionalization, in addition to surfactant and chemical dispersement systems [24]. These methods primarily fall into two categories: mechanical and chemical dispersion. The mechanical dispersion methods, such as ultrasonification, while effective in separating the fibers, can fracture them decreasing their aspect ratio. Chemical methods use surfactants or functionalization to make the fibers less hydrophobic, reducing their tendency to agglomerate. However, many of the chemicals used can digest the fibers causing the fibers to become less effective. The surfactants also often cause bubbles to form in the composite negatively effecting the strength of the material.

Ago et al [12] proposed a dispersion method specifically used for CNF/CNT dispersion in cement-based materials that eliminates the before mentioned drawbacks. In this method, a high-range water reducer (HRWR) is used to create a self-consolidating concrete (SCC). ACI Committee 237 Self-Consolidating Concrete offers the following definition for SCC [25]:

Self-consolidating concrete (SCC) is highly flow able, non-segregating concrete that can spread into place, fill the formwork, and encapsulate the reinforcement without any mechanical consolidation.

SCC is a product of technological advancements in the area of under-water concrete technology where the mixtures must ensure high fluidity and high resistance to washout and segregation. Okamura originally advocated SCC in 1986, and the first success with SCC occurred in 1988 [26]. The use of SCC has gained wide acceptance for savings in labor costs, shortened construction time, a better finish, and an improved work environment [27-30].

Advancement in SCC technology was primarily possible due to the introduction of new chemical admixtures that improved and controlled the SCC rheological properties. Better performing SCC mixes were produced on the advent of melamine, naphthalene, polycarboxylate, and acrylic based HRWR super plasticizers and viscosity modifying agents (VMA).

Gao et al [12] proposed using SCC because acceptable SCC is not only highly flow able, but it is also highly stable and homogeneous on a macro scale. The Prestressed Concrete Institute (PCI) stipulates the following criteria for SCC [26]:

- *Filling ability*: The property that determines how fast SCC flows under its own weight and completely fills intricate spaces with obstacles, such as reinforcement, without losing its stability.
- *Passing ability*: The ability of SCC to pass through congested reinforcement and adhere to it without application of external energy.
- *Stability*: The ability of SCC to remain homogenous by resisting segregation, bleeding, and air popping during transport and placing as well as after placement.

Gao et al [12] studied SCC containing CNF to see if the same effect was present on the nano scale. In GAO Et Al's mixing procedure, HRWR, water, and CNF are mixed in a laboratory-grade blender. Simultaneously, fine aggregate, course aggregate, and cement are combined in a centrifugal mixer. The CNF mixture is then slowly added to the mixer to gain a homogenous mix. The fresh concrete was used to create cylinders that were tested in compression. After the test, pieces of the cylinders were observed under a scanning electron microscope (SEM). The SEM showed significant CNF clumping in specimens made of normal CNF concrete and uniform distribution in SCC containing CNF, as shown in Figs. 2 and 3, respectively.

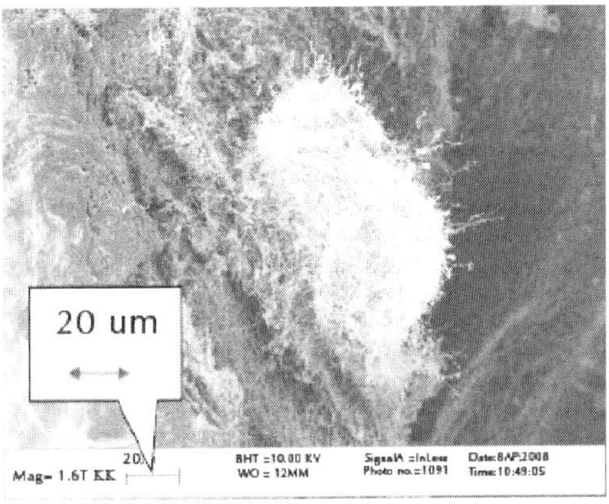

FIGURE 2: Scanning Electron Microscope Image of CNF Clump in Normal Cement (1670x Magnification).

FIGURE 3: Scanning Electron Microscope Image of Well Dispersed CNF in a Uniform Self-Consolidating Cement (9410x Magnification).

STRAIN SENSING ABILITY OF CNT/ CNF CEMENT-BASED MATERIALS

Smart materials are materials that sense their environment and respond to changes in strain, temperature, moisture, pH, and/or electric or magnetic fields. CNT/CNF composites qualify as smart materials since they can be used to measure strain and temperature [4, 13, 32-35]. There are two types of strain sensing, reversible and irreversible. The measurement of irreversible strain allows structural health monitoring, while the sensing of reversible strain permits dynamic load monitoring. Structural health monitoring is the process of implementing a damage detection and characterization strategy for engineering structures. Dynamic load monitoring can detect loads on structures as they are applied and removed in real time. These are important technologies because they gauge the ability of a structure to perform its intended function despite aging, degradation, or disasters. Typically, monitoring reversible strain is more difficult because it can only be monitored in real time. Additionally, reversible strain tends to be smaller than irreversible strain [31].

Strain sensing refers to the ability to measure an electrical or optical response corresponding to a strain. Chen and Chung [31] give the following requirements for a structural sensor:

- Wide strain/stress range of detection (from small strains up to failure)
- Response being reversible upon stimulus removal (necessary for repeated use of the sensor)
- Ease of measuring the response (without the need of expensive peripheral equipment)
- Presence of the sensor having no bad effect on the structural properties of the structure
- Chemical stability and durability
- Low cost

Current commonly used strain sensors include strain gages, fiber optic sensors, and piezoelectric sensors, which all suffer from high cost, poor durability, and the need for expensive peripheral equipment including electronics and lasers. Because of this, the use of sensors in civil structures is uncommon [31]. CNT/CNF composites could become a better option as a strain sensor because however, technology may provide a way to make them more cheaply in the future.

CNT and CNF cement-based materials exhibits properties necessary for reversible strain monitoring and electromagnetic interference (EMI) shielding. Short-fiber composites were found to be a class of strain sensor based on the concept of short electrically conducting fiber pull-out that accompanies slight and reversible crack opening. For a CNT/CNF composite to have strain sensing ability, the fibers must be more conducting than the matrix in which they are embedded, of diameter smaller than the crack length, and well dispersed. Their orientations can be random, and they do not have to touch one another [32, 33]. The electrical conductivity of the fibers enables the DC electrical resistivity of the composites to change in response to strain damage or temperature, allowing sensing [13, 32-35].

CARBON FIBER CEMENT AND MORTAR SELF-SENSING APPLICATIONS

Around the same time that CNT were discovered, reseachers were adding carbon microfibers to cement-based materials and studying their mechanical properties. In 1992 while studying the mechanical properties of carbon microfibers dispersed in mortar, Yang and Chung [35] noted that the electrical resitivity of mortar containing these fibers dramatically decreased by up to several orders of magnitude.

This idea of electrically conducting concrete led to Chen and Chung proposing an intrinsically smart concrete containing carbon microfibers [8]. Chen and Chung prepared mortar cubes containing carbon microfibers and tested them cyclically. They discovered that the electrical resistivity of the concrete increased irreversibly upon compressive loading up to approximately 1/3 the compressive strength of the mortar. After this point, the resistance reversibly increased and decreased upon loading and unloading of the specimens. Chen and Chung concluded that carbon fiber reinforced concrete can serve as a smart structural material. Chen and Chung followed this experiment with a more detailed cyclic experiment on carbon fiber mortar under cyclic loads [31]. After this test, they concluded that the initial irreversible behavior is due to permanent damage associated with the fiber/matrix interface weakening. They attributed the reversible behavior to crack opening with fiber pull-out and crack closing with fibers pushing back in.

CNT are the most conductive fibers presently known and are, therefore, more ideal for electrical applications than their micro-scale counterparts [36, 37]. CNT and CNF are also attractive for use in cement-based composites because of strength and high aspect ratios [6, 16, 17]. Li et al proposed adding MWCNT to mortar for improved mechanical properties [14]. L_i et al confirmed that the flexural and compressive strength of the concrete was enhanced, but they did not study the electrical properties. The same group later studied the electrical volume resistivity of cement paste containing CNT measured using the four-probe method [39]. They applied a cyclic compressive load to a 40.0 mm by 40.0 mm by 160.0 mm (1.575 in. by 1.575 in. by 6.30 in.) rectangular prism made of the material. The fractional change in the volume resistivity oscillated up to approximately 10% with the oscillation of the compressive load.

DAMAGE DETECTION OF CNF CONCRETE COLUMNS

Gao et al expanded the work on self-sensing cement-based materials by studying 152.4 mm by 305 mm (6.00 in. by 12.00 in.) cylindars made of concrete containing CNF [12]. Gao et al crushed the cylinders monotonically and studied the electrical resistance variation. They observed electrical resistance variations up to 80% and concluded that concrete containing CNF can be used for self-structural health monitoring.

Howser et al continued GAO Et Al's work and extended it to a full scale reinforced concrete column containing CNF [4, 12]. A self-consolidating CNF concrete (SCCNFC) column was built and tested under a reversed cyclic load. The structural behavior and the self-sensing ability were examined. The results were compared to the structural and self-sensing ability of a traditional self-consolidating reinforced concrete (SCRC) and a self-consolidating steel fiber concrete (SCSFC) specimen.

All of the columns were 508 mm (20.0 in.) tall with cross-sections of 305 mm by 305 mm (12.00 in. by 12.00 in.). Each specimen contained six #8 (25.4 mm or 1.00 in. diameter) rebar, which corresponded to 3.27% longitudinal steel by volume of concrete. The SCRC and

SCCNFC columns contained #2 stirrups with a spacing of 120.7 mm (4.75 in.) providing transverse reinforcement of 0.287% by volume of concrete. Since the columns were designed to be shear critical, the maximum reinforcement spacing was chosen based on the ACI 318 specifications [25]. See Fig. 4 for the cross-section used for the SCRC and SCCNFC columns. SCSFC column contained no transverse reinforcement, as shown in Fig. 5. Each of the columns was rigidly connected to similar foundations. See Fig. 6 for the elevation view of the SCRC and SCCNFC columns and foundations. The SCSFC column is identical to that shown in Fig. 6, except it does not contain transverse reinforcement. Fig. 7 shows the experimental set-up.

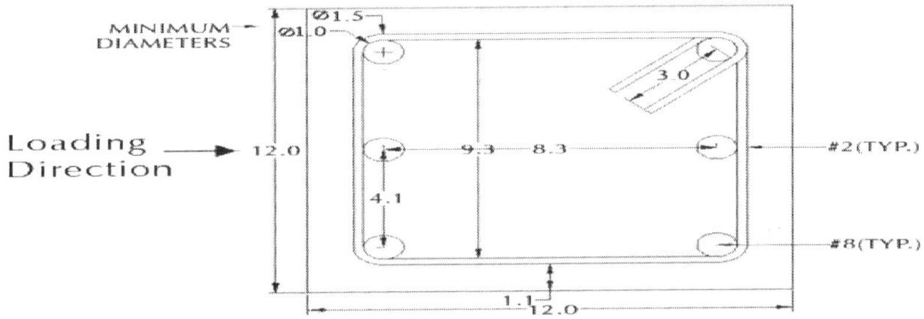

FIGURE 4: Cross-Section of SCRC and SCCNFC Columns (dimensions in inches).

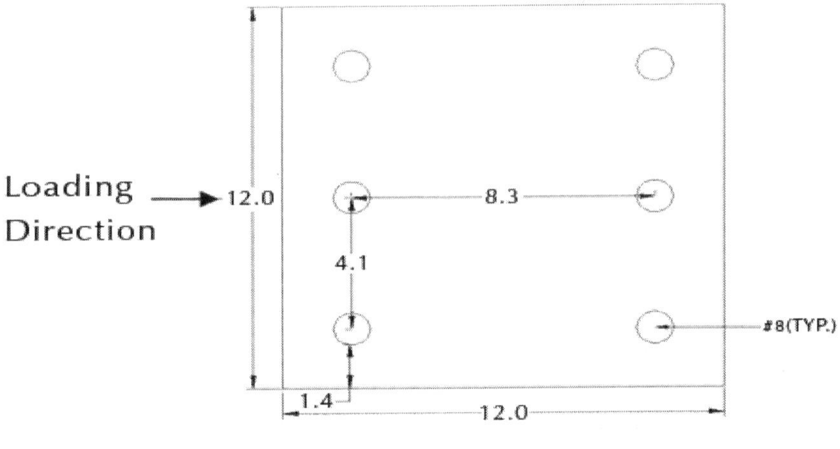

FIGURE 5: Cross-Section of SCSFC Column (dimensions in inches).

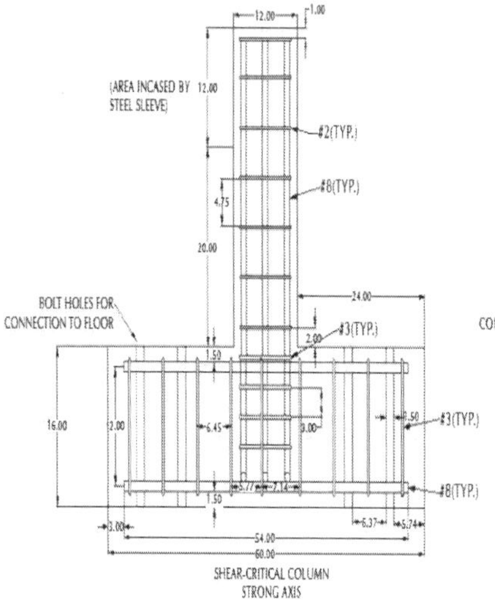

FIGURE 6: Elevation View of the Strong Axis of the Shear-Critical SCRC and SCCNFC Columns and Foundations (dimensions in inches).

FIGURE 7: Experimental Set-Up.

The properties of the materials used for the three mixes were as follows:

1. *Cement*: The cement used in all mixtures was ASTM Type III Portland cement.
2. *Fly Ash*: Class C fly ash was used for the SCSFC mix and Class F fly ash was used for the SCRC mix.
3. *Coarse Aggregate*: Crushed limestore with a maximum diameter of ¾″ was used in the SCCNFC column. River rock with a maximum diameter of ¾″ was used in the other columns.
4. *Fine Aggregate*: Natural river sand with a fineness modulus of 2.71 was used in all mixes.
5. *High Range Water Reducer (HRWR)*: Glenium® 3200HES was used in the SCCNFC column and Glenium® 3400 HES was used in the other columns. Both chemicals were polycarboxylate admixtures from BASF Chemical Co.
6. *Viscosity Modifying Agent (VMA)*: RHEOMAC® VMA 450 was used in the specimens and also supplied by BASF Chemical Co.
7. Steel Fibers: Dramix® ZP305 fibers were used in the SCSFC mix. This was a hooked fiber with a specific gravity of 7.85. The diameter of the fiber is 0.55 mm (0.0217 in.) and the length is 30 mm (1.18 in.) resulting in an aspect ratio of 55.
8. Carbon Nanofibers: Pyrograf Products, Inc. PR-19-XT-LHT-OX fibers were used in this study. The specific gravity of the fibers was 0.0742. The diameter of the fibers was 149 nm (5.87e-6 in.) and the length was 19 µm (7.48e-4 in.) resulting in an aspect ratio of 128.

The mix proportions used for the three columns can be seen in Table 1. One percent fiber by volume was used for both of the fiber columns chosen based on literature review. It was discovered by GAO Et Al that CNF has an optimal dosage of approximately 1% by volume [12] [12]. It was found by many researchers that increased steel fiber increases concrete properties; however, after a percentage of 1% fibers by volume, the concrete becomes increasingly less workable, which could cause problems in construction such as honeycombing [39-41].

TABLE 1: Mix Proportions in kg/m³ (lb/yd³) of Concrete

Material	SCRC Mix	SCSFC Mix	SCCNFC Mix
Cement	446 (752)	446 (752)	457 (771)
Fly Ash (Class C)	-	299 (504)	-
Fly Ash (Class F)	299 (504)	-	-
Fine Aggregate	937 (1580)	937 (1580)	898 (1514)
Coarse Aggregate (Limestone)	-	-	859 (1448)
Coarse Aggregate (River Rock)	491 (827)	491 (827)	-
Water	224 (377)	224 (377)	182 (307)
Glenium ® 3400HES	2.81 (4.73)	2.81 (4.73)	-
Glenium® 7700HES	-	-	2.34 (3.94)
REHEOMAC® VMA 450	5.69 (9.59)	5.69 (9.59)	-
Steel Fibers	-	79.8 (134)	-
Carbon Nanofibers	-	-	3.23 (5.45)

The main goal of testing the SCCNFC column was to prove that concrete containing CNF can be used as a sensor. To test the electrical properties of the concrete, wire meshes were constructed and embedded in each of the columns. The wire meshes were made of 12.7 mm (1/2 in.) hardware cloth with 14 gauge copper wire soldered to it. The wire extended outside of the column. The four probe method for measuring resistance was implemented, and the meshes were placed in the column as shown in Fig. 8. A power supply was attached to the top mesh that provided a current of approximately 31 V DC. An ammeter was attached to the bottom mesh and connected back to the power supply to complete a circuit. The current measured by the ammeter was recorded continuously during the tests by hand. Additional voltmeters were attached to the two middle meshes on both the north and south sides of the column to measure voltage. The voltage readings were also recorded continuously throughout the test.

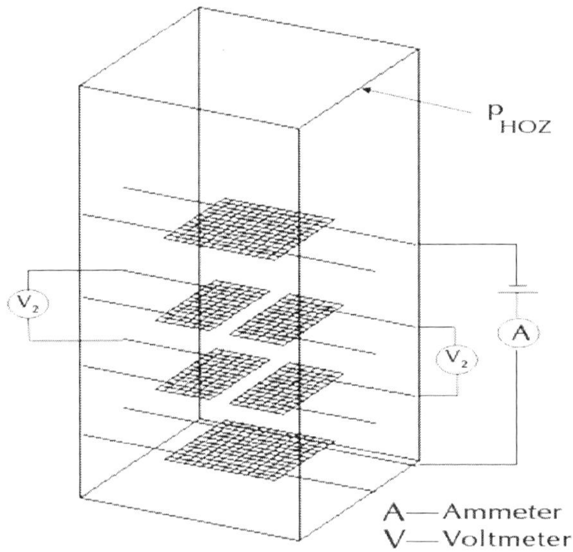

FIGURE 8: Four Probe Method of Resistance Measurement.

The first step of the load program was to apply an axial load that would remain constant during the course of the test. The axial load equaled one-tenth of each of the columns calculated axial capacity. The axial capacity is dependent on the compressive strength of the concrete, so the axial force for each specimen varied.

After the application of the axial load, a reversed-cyclic load was added using a 649 kN (146 kip) capacity actuator. The intended load path was to use force control to complete two cycles each of ±89 kN (20 k), ±178 kN (40 k), and ±267 kN (60 k). A positive force denotes a push by the actuator while a negative force represents a pull. At the point of longitudinal steel yielding, the test switched to displacement control and completed two cycles each at a displacement ductility of 2, 3, 4, etc. Once failure occurred, a descending branch on the load versus displacement curve was obtained in displacement control mode.

The load path followed for the SCRC column specimen can be seen in Fig. 9 with the first cracks, switch to displacement control and failure marked. The first crack on the south side of the column occurred at -178 kN (-40 k). The first shear crack formed on the column during the first -178 kN (-40 k) cycle at -178 kN (-40 k) on the west side. The column failed in shear and crushing of concrete at 276 kN

(62 k). The west side of the column exhibited crushing of the concrete struts with large shear cracks. The east side exhibited local crushing at the actuator connection. The maximum displacement at the top of the column (drift) was 12.7 mm (0.50 in.).

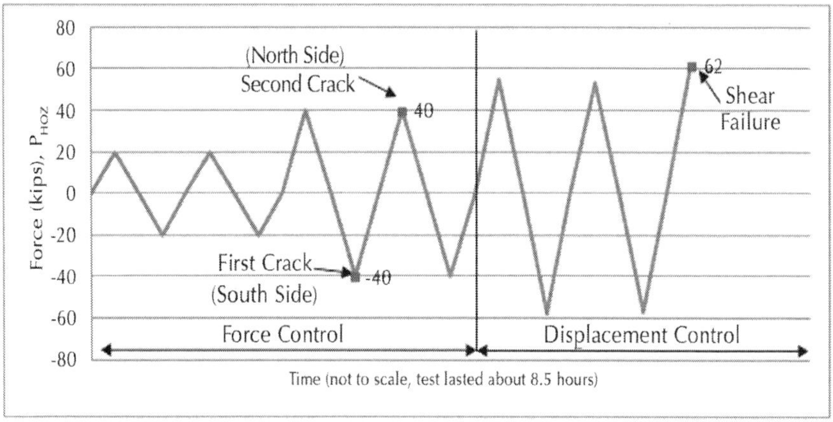

FIGURE 9: SCRC column load path.

The load path followed for the SCSFC column can be seen in Fig. 10 with the first cracks and failure marked. The first shear and flexural cracks formed on the column during the second 178 kN (40 k) cycle at 178 kN (40 k) on the west and north sides, respectively. The second flexural crack formed on the south side during the second -178 kN (-40 k) cycle at -178 kN (-40 k). The column failed suddenly in shear and crushing at 347 kN (78.0 k) on the west side of the column before the rebar yielded. The maximum displacement was 8.38 mm (0.33 in.).

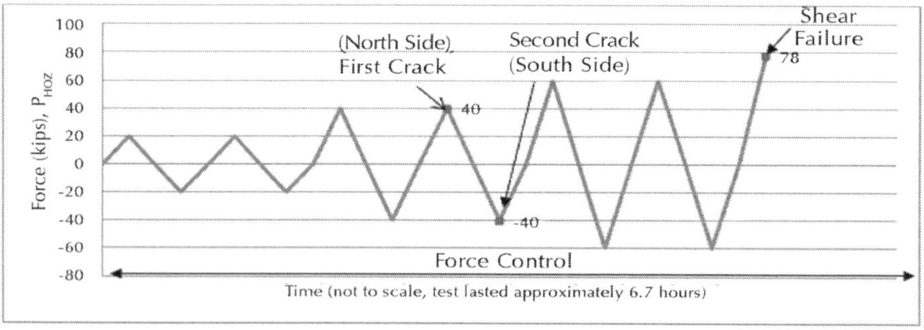

FIGURE 10: SCSFC column load path.

The actual load path followed for the SCCNFC column can be seen in Fig. 11. The pump shut down during the test, and the actuator unloaded during the fifth cycle of the test. The pump was turned back on and the test resumed. The first flexural crack formed on the column at 160 kN (36 k) on the east, west and north sides. The second flexural crack formed on the east, west and south sides at a load of -158 kN (-35.6 k). The column failed in the combined modes of shear and concrete crushing due to flexure at 298 kN (67 k) on the west side of the column. The maximum displacement was 10.16 mm (0.4 in.).

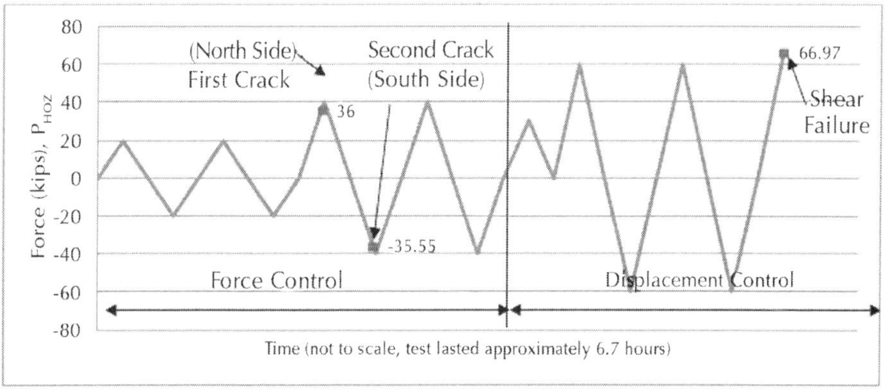

FIGURE 11: SCCNFC column load path.

During each of the column tests, the electrical resistance was determined to check the self-sensing ability of the concrete. The electrical readings showed a great correlation between the peaks in the applied horizontal force, strain, and resistance plots for the SCCNFC column but little correlation between the resistance plots and the force or strain plots for the SCRC or SCSFC column. Fig. 12 shows the relationship between the SCRC column's horizontal force, LVDT strain, and electrical resistance versus time on the north side of the column. There is no relationship between the peaks and valleys in the electrical resistance and the load or strain on the north side of the column.

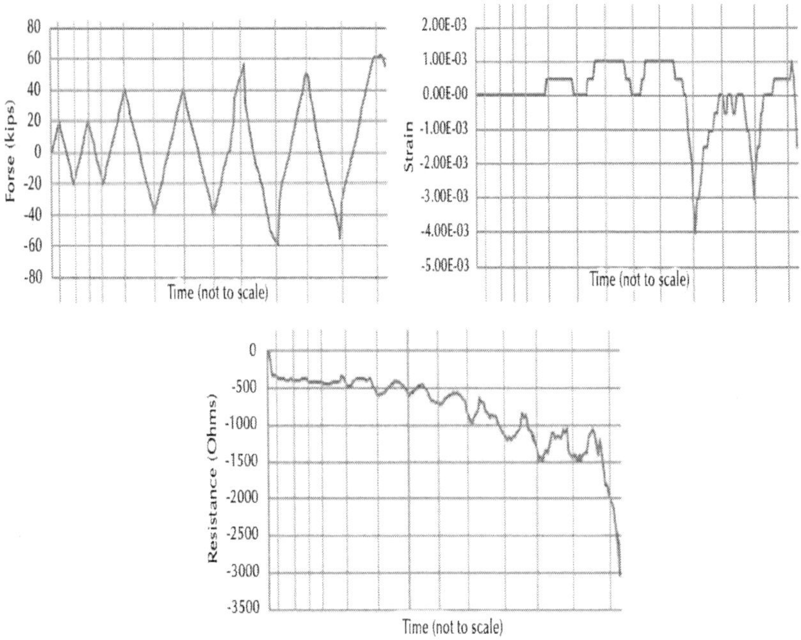

FIGURE 12: SCRC Column Comparison of Horizontal Force, LVDT Strain and Electrical Resistance on North Side.

Fig. 13 shows the SCSFC column's force, strain, and resistance versus time on the north and south sides of the column, respectively. As shown by the grey vertical lines, there is not a relationship between the peaks and valleys in the resistance and load or strain until major cracking began to occur. After major cracking began to occur, the peaks and valleys in the electrical resistance began to correspond with the load and strain peaks and valleys. This point is shown by the dashed line in Fig. 13.

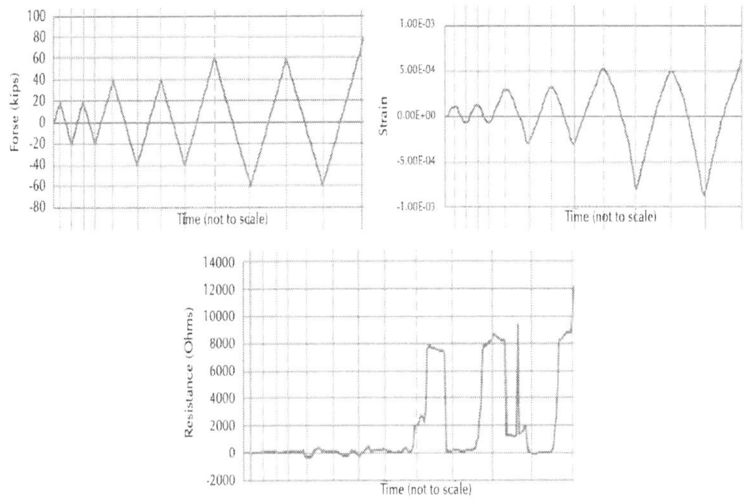

FIGURE 13: SCRC Column Comparison of Horizontal Force, LVDT Strain and Electrical Resistance on North Side.

Fig. 14 shows relationship between the SCCNFC column's horizontal load, LVDT strain and electrical resistance versus time on the north side of column. As shown by the vertical lines in the grid, there is very good correlation between the force, strain and resistance. On the north side of the column, the peaks and valleys matched up until the point that the column was greatly damaged.

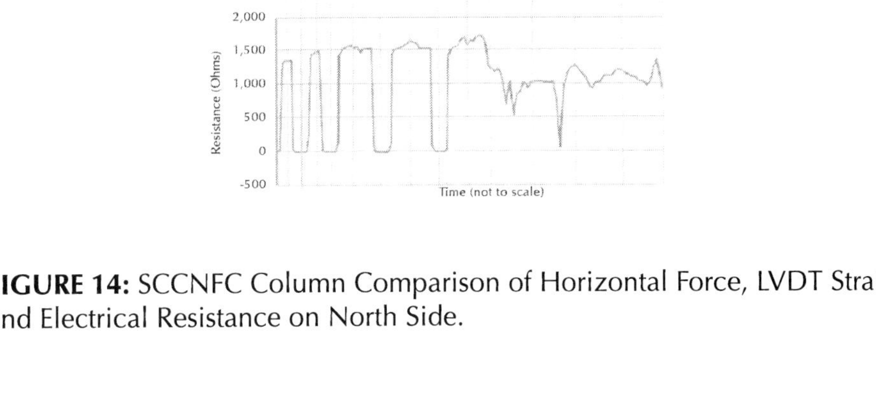

FIGURE 14: SCCNFC Column Comparison of Horizontal Force, LVDT Strain and Electrical Resistance on North Side.

Because of the strong correlation found between the horizontal load, LVDT strain, and electrical resistance verses time graphs for the SCCNFC column, the electrical resistance variation (ERV) was calculated and compared to the deflection at the top of the column. ERV is the measured electrical resistance minus the initial electrical resistance quantity divided by the initial electrical resistance. Fig. 15 shows the relationship between the ERV and deflection at the top of the column for the first five cycles of the test. It is obvious from Fig. 15 that the column shows major damage at approximately a deflection of 2.03 mm (0.08 in.). This corresponds to the steel yielding in the SCCNFC column. This proves that SCCNFC can be used as a type of self-structural health monitoring system.

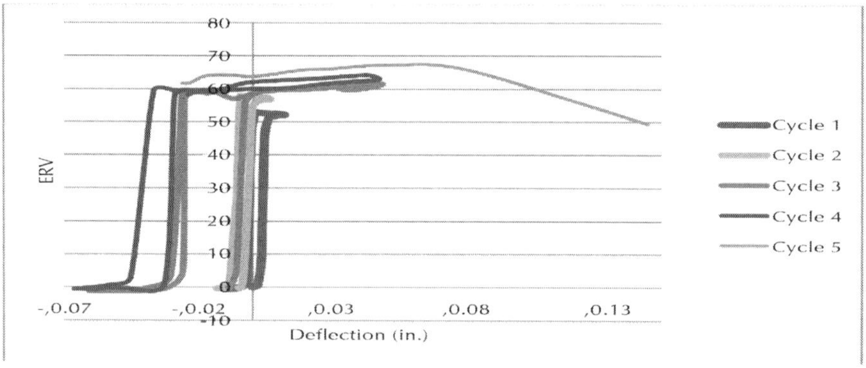

FIGURE 15: SCCNFC Column ERV versus Horizontal Deflection.

CONCLUSIONS

Self-consolidating carbon Nano fiber concrete (SCCNFC) follows the definition for nanotechnology set forth by the National Science Foundation and National Nanotechnology Initiative [2]. The size range of the carbon Nano fibers (CNF) is approximately 100 nanometers, the SCCNFC is able to measure damage in the composite, and the CNF have properties that are specific to the Nano scale.

Well-dispersed CNF improves the strength and stiffness of concrete. Excess concentration leads to poorly dispersed CNF clumps inside the concrete and has a negative effect on both strength and electrical sensitivity. Highly workable and stable self-consolidation concrete (SCC) can maintain its workability and stability with the addition of fibers. SCC greatly increases the dispersion of carbon Nano fibers (CNF) [12].

As proven by Gao et al [12] and Howser et al [4], SCCNFC can be used as a reversible strain sensor. In Hawser et al's test [4], the peaks and valleys in the electrical resistance readings of the SCCNFC match the peaks and valleys of the applied force and the strain in the concrete. While the peaks and valleys in the electrical resistance readings of the self-consolidating reinforced concrete and self-consolidating steel fiber concrete specimens occasionally matched, there was not enough correspondence to safely assume that these concretes could be used as a reversible strain sensor. It was concluded that when an appropriate dosage of CNF is used, SCCNFC can be used for self-structural health monitoring.

REFERENCES

1. Feynman, R. There's Plenty of Room at the Bottom. Engineering and Science 1960; 23, 22-36.
2. Roco, M. C. National Nanotechnology Initiative - Past, Present, Future. In W. A. Goddard, D. Brenner, S. E. Lyshevski, & G. J. Iafrate (Eds.), Handbook of Nanoscience, Engineering, and Technology (2nd ed.). Boca Raton, FL: CRC Press 2007; 3.1-3.26.
3. Narayan, R. J., Kumta, P. N., Sfeir, C., Lee, D.-H., Choi, D., Olton, D. Nanostructured ceramics in medical devices: Applications and prospects. Journal of the Minerals, Metals and Materials Society 2004; 56 (10), 38-43.
4. Howser, R. N., Dhonde, H. B., Mo, Y. L. Self-sensing of carbon nanofiber concrete columns subjected to reversed cyclic loading. Smart Materials and Structures 2011; 20 (8).
5. Bartos, P. Nanotechnology in Construction: A Roadmap for Development. American Concrete Institute Special Publication 2008; 254, 1-14.
6. Sanchez, F., Sobolev, K. Nanotechnology in concrete - A review. Construction and Building Materials 2010; 24, 2060-2071.

7. Shah, S. P., Naaman, A. E. Mechanical Properties of Glass and Steel Fiber Reinforced Mortar. ACI Journal 1976; 73 (1), 50-53.
8. Chen, P.-W., Chung, D. D. Concrete reinforced with up to 0.2 vol% of short carbon fibres. Composites 1993; 24 (1), 33-52.
9. Li, H., Xiao, H.-g., Yuan, J., Ou, J. Microstructure of cement mortar with nano-particles. Composites Part B: Engineering 2004; 35 (2), 185-189.
10. Li, H., Zhang, M.-h., Ou, J.-p. Abrasion resistance on concrete containing nano-particles for pavement. Wear 2006; 260 (11-12), 1262-1266.
11. Li, H., Zhang, M.-h., Ou, J.-p. Flexural fatigue performance of concrete containing nano-particles for pavement. International Journal of Fatigue 2007; 29 (7), 1292-1301.
12. Gao, D., Strum, M., Mo, Y. L. Electrical resistance of carbon-nanofiber concrete. Smart Materials and Structures 2009; 18 (9), 1-7.
13. Iijima, S. Helical microtubules of graphitic carbon. Nature 1991, 354, 56-58.
14. Li, G. Y., Wang, P. M., Zhao, X. Mechanical behavior and microstructre of cement composites incorporating surface-treated multi-walled carbon nanotubes. Carbon 2005; 43, 1239-1245.
15. Salvetat, J. P., Bonard, J. M., Thomson, N. H., Kulik, A. J., Forro, L., Benoit, W., et al. Mechanical properites of carbon naonotubes. Applied Physics A 1999; 69, 255-260.
16. Makar, J. M., Beaudoin, J. J. Carbon nanotubes and their application in the construction industry. 1st International Symposium on Nanotechnology in Construction, Paisley, Scotland 2003; 331-341.
17. Kang, I., Heung, Y. Y., Kim, J. H., Lee, J. W., Gollapudi, R., Subramaniam, S., et al. Introduction to carbon nanotube and nanofiber smart materials. Composites: Part B 2006; 382-394.
18. Coleman, N. N., Khan, U., Blau, W. J., Gun'ko, Y. K. Small but strong: A review of the mechanical properties of carbon nanotube-polymer composites. Carbon 2006; 44, 1624-1652.
19. Makar, J. M., Margeson, J. C., Luh, J. Carbon nanotube/cement composites-early results and potential applications. 3rd International Conference on Construction Materials: Performance,

Innovation and Structural Implications. Vancouver, B.C.: NRC Institute for Research in Construction; National Research Council Canada 2005; 1-10.

20. Chung, D. D. Dispersion of Short Fibers in Cement. Journal of Materials in Civil Engineering, ASCE 2005; 17 (4), 379-383.
21. Tzeng, Y., Huang, T. S., Chen, Y. C., Liu, C., Liu, Y. K. Hydration Properties of Carbon Nanotubes and Their Effects on Electrical and Biosensor Applications. New Diamond and Frontier: Carbon Technology 2004; 14 (3), 193-201.
22. Baughman, R. H., Cui, C., Zakhidov, A. A., Iqbal, Z., Barisci, J. N., Spinks, G. M., et al. Carbon Nanotube Actuators. Science 1999; 284 (5418), 1340-1344.
23. Baughman, R. H., Zakhldov, A. A., de Heer, W. A. Carbon Nanotubes--the Route Towards Applications. Science 2002; 297, 787-792
24. Hilding, J., Grulke, E. A., Zhang, Z. G., & Lockwood, F. Dispersion of Carbon Nanotubes in Liquids. Journal of Dispersion Science and Technology 2003; 24 (1), 1-41.
25. ACI 318. Building Code Requirements for Structural Concrete. Farmington Hills, Michigan: American Concrete Institute; 2008.
26. PCI, TR-6-03. Interim Guidelines for the Use of Self-Consolidating Concrete in Precast/Prestressed Concrete Institute Member Plants. Chicago: Precast/Prestressed Concrete Institute; 2003.
27. Gaimster, R., Foord, C. Self-compacting concrete. Concrete 2000; 34, 23-25.
28. Khayat, K. H., Hu, C., Monty, H. Stability of Self-Consolidating Concrete, Advantages, and Potential Applications. Proceedings of First International RILEM Symposium on Self-Compacting Concrete. Stockholm, Sweden 1999; 143-152.
29. Okamura, H., Ozawa, K. Mix-Design for Self-Compacting Concrete. Concrete Library of the Japanese Society of Civil Engineers 1995; 107-120.
30. Tanaka, K., Sato, K., Watanabe, S., Arima, I., Seunaga, K. Development and Utilization of High-Performance Concrete for the Construction of the Akashi Kaikyo Bridge. High Performance Concrete in Severe Environments, ACI SP-140 1993; 25-51.

31. Chen, P.-W., Chung, D. D. Concrete as a new strain/stress sensor. Composites Part B: Engineering 1996; 27 (1), 11-23.
32. Chung, D. D. Strain sensors based on the electrical resistance change accompanying the reversible pull-out of conducting short fibers in a less conducting matrix. Smart Materials and Structures 1995; 4 (1), 59-61.
33. Chung, D. D. Cement-matrix composites for smart structures. Smart Materials and Structures 2000; 9 (4), 389-401.
34. Xiao, H., Lan, C., Ji, X., Li, H. Mechanical and sensing properties of structural materials with nanophase materials. Pacific Scientific Journal 2003; 5, 11-17.
35. Yang, X., Chung, D. D. Latex-modified cement mortar reinforced by short carbon fibres. Composites 1992; 23 (6), 453-460.
36. Thess, A., Lee, R., Nikolaev, P., Dai, H., Petit, P., Robert, J., et al. Crystalline Ropes of Metallic Carbon Nanotubes. Science 1996; 273, 483-487.
37. Wei, B. Q., Vajtai, R., Ayayan, P. M. Reliability and current carrying capactiy of carbon nanotubes. Applied Physics Letters 2001; 79 (8), 1172-1174.
38. Li, G. Y., Wang, P. M., Zhao, X. Pressure-sensitive properties and microstructure of carbon nanotube reinforced cement composites. Cement & Concrete Composites 2007; 29, 377-382.
39. Narayan, R. J., Kumta, P. N., Sfeir, C., Lee, D.-H., Choi, D., Olton, D. Nanostructured ceramics in medical devices: Applications and prospects. Journal of the Minerals, Metals and Materials Society 2004; 56 (10), 38-43.
40. Padmarajaiah, S. K., Ramaswamy, A. Comparative Study on Flexural Response of Full and Partial Depth Fiber-Reinforced High-Strength Concrete. Journal of Materials in Civil Engineering 2002; 14 (2), 130-136.
41. Aoude, H., Cook, W. D., Mitchell, D. (2009). Behavior of Columns Constructed with Fibers and Self-Consolidating Concrete. ACI Structural Journal 2009; 106 (3), 349-357.

Chapter 3

Effect of Alternative De-icers on the Corrosion Resistance of Reinforced Concrete Bridges and Highway Structures

S. O. Nwaubani[1] and A. Katsanos[2]

Department of Engineering and the Built Environment, Anglia Ruskin University, Cambridge and Chelmsford, Essex, UK

Department of Civil Engineering, School of Engineering, University of Surrey, Guildford,

Surrey, UK

INTRODUCTION

Prolonged periods of snowfall in countries with advanced infrastructure and transport systems have rendered the use of de-icing agents to a common occurrence on roads and highway structures. They are necessary in order to maintain a good level of service with respect to the transport systems, thus avoiding traffic jams and disruptions, but also to provide a high level of road safety. Today, chloride-based products, such as rock salt, are the most commonly encountered de-icers as they are easy to apply and store but mostly because they efficiently melt ice at an affordable price [1]. However, their widespread use over a long period has left the construction industry and the engineering community with a grave problem regarding the durability of highway reinforced concrete bridges and multi-storey parking structures [2], due mainly to the fact that they cause corrosion of the reinforcement and steel components [2, 3].

For reinforced concrete bridges, in particular, the chlorides reach the reinforcement through leaking joints and spray from passing traffic. The corrosion product is known to occupy 2 to 6 times the volume of the metal from which it originates and this results to spalling of the concrete cover [2]. This process has a multitude of implications for the durability of concrete structures as it may imply section loss of the reinforcement and compromise of their load bearing capacity. Loss of the protection provided by the concrete cover further promotes and accelerates corrosion, thus leading to structural deficiencies and reduction of the life cycle of the structure. In the US alone, there are 20-25 year-old bridges, which are already facing this type of problems due to heavy de-icer applications and inadequate maintenance [1]. Furthermore, chloride de-icers have also been reported to have a negative impact on the natural environment [4, 1], and accelerate the corrosion of vehicle parts and roadside objects [5, 1]. These factors have inevitably raised concerns regarding the sustainability of current de-icing practices.

COST OF CORROSION IN GENERAL

The annual cost of corrosion consists of both direct costs and indirect costs. The direct costs related to corrosion are made up of two main components:

a. The costs of design, manufacturing, and construction such as; use of new materials such as stainless steel to replace carbon steel; use of novel technologies to achieve increased wall thickness for corrosion allowance; innovations in the mitigation and/or prevention of corrosion, such as coatings, sealants, corrosion inhibitors, cathodic protection, and cost of labour and equipment.

b. The direct and indirect cost of management relating to the inspection, repair, maintenance, replacement of corroded parts, rehabilitation, and loss of productive time in corrosiondamaged structures.

Using highway bridges as an example, the optimized contribution of each of the contributing components is calculated through life cycle cost analysis and characterized by the annualized value. Relevant surveys carried out worldwide has highlighted the heavy cost of maintaining, repairing, and replacing concrete structures due to corrosion. In the United states a Corrosion Costs Study carried out by the National Association of Corrosion Engineers (NACE), revealed the direct cost of corrosion in the U.S. was $276 Billion in 1998, approximately 3.1 % of GDP. The indirect cost of corrosion was also estimated to be at least equal to the direct cost giving a total cost of 552 Billion. The direct and indirect costs increased to $468 Billion in 2011, giving a total cost due to corrosion of $936 Billion in that year. In fact, a significant milestone in the effect of corrosion on the U.S. economy was reached this year (2013), when the total cost of corrosion in the US exceeded $1 trillion annually for the first time (based on estimates of GDP from http://forecasts.org/gdp.htm). This represents 6.2% of the GDP.

Table 1. Result from (NACE Corrosion Costs Study)

Year of Study			
	1998	2011	2013
Direct corrosion cost	$2 76Billion	$468 Billion	>$500 Billion
Indirect corrosion cost	$275 Billion	$468 Billion	>$500 Billion
Total Cost	$552 Billion	$936 Billion	>$1 Trillion

An earlier study by Koch, et al [6], established the cost of corrosion to the Government sector, the Production and Manufacturing sector, Utilities sector, Infrastructure and Transportation sector. Ofthese,the costof corrosioninthe infrastructure categorywas estimatedat$22.6billion, which was 16.4 % of the total cost of the sector categories examined in the study (see Figure 1).

The infrastructure category is divided into the following industry sectors: (a) highway bridges, (b) gas and liquid transmission pipelines, (c) waterways and ports, (e) hazardous materials storage, (f) airports, and (g) railroads.

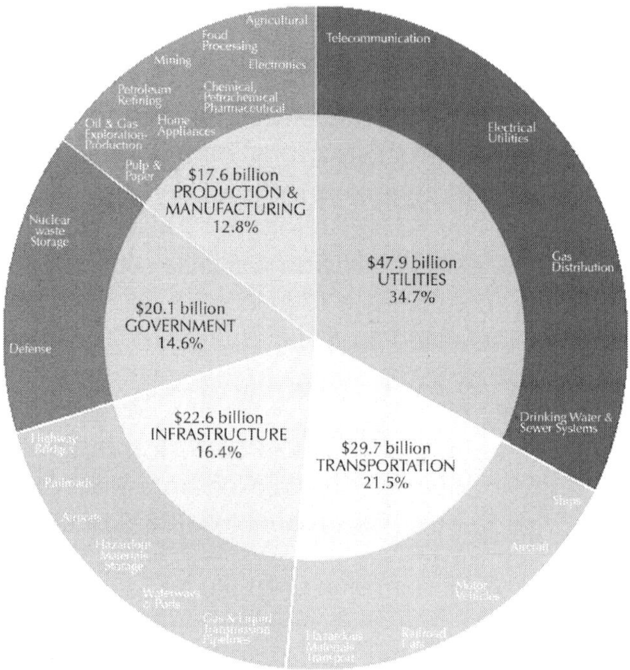

Figure 1: NACE Corrosion Costs Study [6].

Facts about Reinforced Concrete Bridge Corrosion

There are about three quarters of a million conventional reinforced concrete bridges in the USA. According to the latest Federal Highway Administration's National Bridge Inventory, at least 59% of these are reinforced concrete structures [6] and over 30% of the bridges are structurally deficient or functionally obsolete. The annual direct cost of corrosion for highway bridges is estimated to be $13.6 billion. In the United States, the direct annual cost of repair and replacement of deficient concrete bridges has been estimated to be around US$8.3 billion [7, 8], or 0.095% of the 1998 GDP of that country [9]. A similar survey conducted in Japan [10], estimated the corresponding expenditure to be ¥177.4 billion, or 0.059% of the 1999 GDP of that country [11]. In Britain the total estimated cost for the rehabilitation and repair of deficient concrete bridges was found to be £616.5 million spread over a period of ten years after 1989 when the report was published [12]. This would amount to about 0.013% of the 1989 GDP of the UK according to the Office for National Statistics and the figures does not include costs relating to damage on roadside objects, vehicles, the loss of floral and faunal species and surface water quality, the adverse effects on soil structure and the migration of sodium and chloride ions into water supplies [1]. It is also estimated that indirect costs, such as traffic delays and subsequent lost productivity, can amount to ten times the value of direct costs [1].

Cost of corrosion studies have been undertaken by several other countries, not cited above, including, Australia, Kuwait, Germany, Finland, Sweden, India, and China. The studies have ranged from formal and extensive efforts to informal and modest efforts. The common finding of these studies was that the annual corrosion costs ranged from approximately 1 to 5 percent of the Gross National Product (GNP) of each nation [5]. Furthermore, the cases cited above are mainly before the millennium, however an important point to note is that the cost of remediation of corrosion-damaged structures is ever increasing, even in this decade and millennium, with many developed countries spending an ever greater proportions of their GDP for maintenance and repair of corrosion-damaged infrastructure (see Table 1 below).

FACTS ABOUT DE-ICERS

Chloride-Based De-Icers

Due to extensive research on ice-melting chemicals in recent years and their many different applications, there is currently a multitude of solid and liquid road de-icing and anti-icing products on the market. The most common type of de-icers used today is the chloride-based de-icers and in particular rock salt, composed primarily by sodium chloride (NaCl). A figure typical of the extensiveness of its use is the 10 million tons of salt used in the US every year [13, 14]. It is used so widely because of its low price, ease of application, and documented effectiveness. Although chloride-based de-icers perform well with respect to ice melting, their aforementioned impact on the environment and infrastructure has led the industry to pursue different methods of winter operations, which could limit or eliminate their use [1]. In this context, the emerging, acetate-based de-icing chemicals have been chosen for investigation both on-site and in the laboratory by many transportation departments and research establishments especially in the US ([1], as there is evidence that they could provide a more sustainable solution towards winter operations, despite their, currently, high cost. In what follows, this chapter will explore in greater depth the properties of the two common Acetatebased de-icers: CMA (calcium magnesium acetate), and NAAC (sodium acetate)

Acetate-Based De-Icers

Calcium Magnesium Acetate (CMA)

Recognition of the negative effects of chloride-based de-icers has generated research into alternative de-icers, especially in the United States. Calcium magnesium acetate (CMA) is one of the chemicals considered for this purpose. CMA has been investigated for its ice-melting characteristics [15], its effects on concrete [16,17,18,19,20] and various metals including reinforcing steel [21,7,8], its environmental impact as well as other aspects. Almost all of these studies have found CMA to be effective and benign. The drawback however, is that its price currently ranges between 20 to 30 times the cost of rock salt [22, 1]. A report carried out following a request by the US congress [1], identified many benefits which could arise from the replacement of rock salt by CMA. Nevertheless, it highlighted the fact that it was doubtful whether a complete substitution would provide the expected financial benefits; a problem compounded by the fact that much of the data were not easy to quantify. Subsequent studies, by Katsanos and

Nwaubani [23], along with data resulting from field experience [24, 25, 26, 27, 28] have all reinforced the view that the use of CMA is a safe and sustainable option.

Table 2. Cost of Corrosion

Country/ Project	GOP	Annual Cost of Corrosion	Source of information
USA(1998)	$276 Billion	3.1 5% of GOP	Corrosion Costs and Preventiw Strategies in the United States', Report FHWA·R0-01·156· http:// wVliW.nace.org
USA(2007)	$429	$13,840 billion	NACE figures: http://eve nts.nace.org/ publicaffairs/
	Billion		cocorrindex.asp
USA(201 3)	$16.535 Trillion	>1 trillion (6.2% GOP)	Bureau of Economic Analysis, Department of Commerce. July 31,2 013 Inco*me* and Product Accounts Gross Domestic Product, second quarter 2013 (advance estimate); http:/ hMMN.bea.gov/ *newsre* lease s/national/gdp /2013/. Retrieved August 1, 2013.
Canada(2005) Roads, Sidewalks, Bridges Toronto (Canada)	$992 Billion	$110 million spent in 2005 ... with a backlog of $235 million deferred due to budget constraints	K. McGran's article "On the road to ruin?", Toronto Star, February 5, 2005, pB4-B5.
Canada(2012)	$1.819 Trillion	2-4% of GDP	Bureau of Economic Analysis, Department of Commerce. July 31, 2013
Japan (1997)	$ 4.324 Trillion	0.8-1.0 % of GNP (1997 estimate of direct corrosion costs)	National Institute for Materials Science (Japan).
Japan(2012)	$5.964 Trillion	2-4% of GDP	Bureau of Economic Analysis, Department of Commerce. July 31, 2013

Australia(2008)		A$1-5 billion Australian dollars [~ 2% of GDP	CSIRO-Re: Holistic Approach to Corrosion control, January 2008 web page.
Australia(2009)	$920 Billion	$70.6 billion	
Australia (2012)		$1,542 Trillion 2-4% of GDP	Bureau of Economic Analysis, Department of Commerce. July 31, 2013
United Kingdom (2007), Corrosion assessment, mitigation and repair of the old Severn Bridge.		£23 million	BBC News, March 7, 2007. (On-line Edition)
United Kingdom (2008)	$2.279 Trillion	$70.6 billion	GDP figures: http://www.economywatch.com/
United Kingdom (2012)	$2,441 Trillion	3-4% GDP	Structure and Infrastructure Engineering Maintenance, Management, Life-Cycle Design and Performance, Taylor & francis Volume 9, Issue 12, 2013, ISBN 1573-2479.

Sodium Acetate (NAAC)

Another acetate de-icer, sodium acetate (NAAC), which has a similar cost to CMA, is currently used principally as an airport runway de-icer. It has been used been used for road de-icing in the towns of Aspen and Snowmass in the USA and has reportedly performed exceptionally well at very low temperatures [22]. NAAC is exothermic and this property makes it more effective in ice-melting as it stays active at lower temperatures than CMA [22]. Its potential use as a highway de-icer has not been fully researched, little or no few field experience has been documented regarding the effects of NAAC on corrosion behaviour of embedded steel reinforcement although it is commonly regarded as a non-corrosive de-icer [29, 30]. This is possibly due to concerns regarding its potential effects on surface waters. There are fears that, if used, it could increase the BOD of surface waters, more than other acetate de-icers, but less than urea [22]. If NAAC proves to be efficient in minimising reinforcement corrosion, and can be demonstrated to have an acceptable environmental impact, then it could provide a viable solution for use on corrosion prone structures in areas where temperatures fall below the active range of CMA.

De-icing chemicals are incapable of melting snow and ice in their dry (solid) state. They must first attract or encounter moisture to form brine (a chemical/water solution). The brine then penetrates down through the ice and snow until it reaches the pavement, then spreads outwards melting and undercutting the ice and snow for mechanical removal. Although the same chemicals behaviour are involved, there are remarkable differences in how de-icers work. A basic reason for these differences in de-icing is that some chemicals take longer to go into solution before they can begin penetrating ice and snow. Another reason for de-icing action is heat liberation. As they dissolve, some chemicals liberate heat due to exothermic reaction. In addition, pellet de-icers have been found to be more efficient than flake de-icers [24]. Generally, the performance of a de-icer is determined by its ability to melt, penetrate, undercut, and break the bond between the ice and snow. Two of the factors that determine the ability of a de-icer to perform these functions are the eutectic and effective temperatures of the de-icer. From basic physics, it is known that the eutectic temperature is the theoretical lowest temperature to which the de-icer can suppress the freezing point of water. However, this temperature can only be reached at the ideal percent concentration of the de-icer in water. All other concentrations will have higher freezing points [13]. Because de-icers become increasingly diluted as they melt ice, they remain at their ideal concentration only briefly. Consequently, the eutectic temperature has little bearing on real-life de-icing conditions. The effective temperature, which is the lowest temperature at which each de-icer remains effective, is a much more realistic measure of the de-icing capability of these chemicals [22]. Site experience gained by the use of acetate de-icers over recent years, has provided a clearer picture with respect to the way that acetate de-icers work and their effectiveness in comparison to chloride de-icers. Table 3 displays the eutectic and effective temperatures of the de-icers under consideration. It becomes obvious that the effective temperature of sodium acetate is much lower due to the fact that it possesses exothermic properties.

Table 3. Eutectic and Effective Temperatures of selected de-icers [13,14]

DE-ICER	EUTECTIC CONCENTRATION (%)	EUTECTIC TEMPERATURE (C°)	EFFECTIVE TEMPERATURE (C°)
CMA (anhydrous calcium magnesium acetate)	32.60	-27.78	-6.67
NAAC (anhydrous sodium acetate)	27.00	-21.67	-15.00
Sodium Chloride	23.00	-21.12	-9.44

In another study Shenk [31], observed that the slower rate of ice melting arising from the use of CMA was due primarily, to the tendency of the CMA crystals to be buoyed up in their solution. However, the study revealed that CMA, is in many ways superior to rock salt as a road de-icer, due to its ability to melt ice at lower temperatures, and a marked heat of solution of the anhydrous CMA.

Site Experience on the Effectiveness of CMA and Sodium Acetate Deicers

Field evaluation of the performance of CMA was studied in two similar articles from the late 1980's [24, 25]. They discussed experience from CMA applications at different locations. A common finding from those field studies is that CMA works differently from Sodium chloride. CMA prevents snow pack when applied early in a storm, forming a fluffier, drier snow-pack easy to remove by ploughing or by passing traffic, while Sodium chloride produces a wetter and heavier snow-pack. Therefore, timing was crucial in the case of CMA application

MI92 MI93.The downside to early application is that, if done before a snow storm, it runs the risk of being blown away by traffic due to its low density. If applied late it may be slower acting, taking around 20 minutes longer to act compared to rock salt [25].

CMA also seems to work at lower temperatures than rock salt, as tendencies to re-freeze after temperature drops seem less, according to the observations made at the Zilwaukee Bridge [27, 28]. It also lasts longer and it sometimes has to be applied once during a 12-hour period compared to two or three for rock salt. This could also be due to the residual effect that CMA seems to have on the road surface as liquefied CMA and pellets stick on the road surface minimising re-freezing and early morning deck icing [27]. Clearly, CMA acts in a different manner and this explains the initial tendency of early users and researchers to believe that a larger amount of CMA than rock salt would be required to achieve equal de-icing performance. A study conducted in 1993 [26] attributes the CMA-snow-pack interaction to the ability of the calcium acetate and the magnesium acetate to dissociate from the CMA compound and form independent hydrates. Similar studies have not, however been conducted for sodium acetate, and as a result we know little about its corrosiveness with respect to concrete reinforcement. Nonetheless, Sodium Acetate complies with SAE standard AMS 1431 regarding its suitability for use on runways, and is deemed non-corrosive to metal parts of aeroplanes.

Effects of Acetate De-Icers on Corrosion of Steel Reinforcement

As the cost of maintaining, repairing and replacing concrete structures such as highway bridges or multi-storey parking buildings soared, academics and the industry were urged to investigate the corrosiveness of non-chloride-based de-icers. The chemical identified as the most likely substitute, as mentioned above, was CMA, and numerous projects have so far been undertaken in order to assess its effects on concrete and on steel reinforcement.

With respect to the corrosiveness of CMA, a study conducted by the Michigan transportation commission [7,8] deduced that prestressing steel strands immersed in NaCl solutions experienced 4 to 15 times more corrosion than identical specimens immersed in CMA solutions designed to have the same freezing point. The study also concluded that mixtures of the two materials produced results comparable to the ones produced by the CMA solutions down to at least a 0.46 CMA/NaCl weight ratio. However, this last claim is dismissed by a study which measured corrosion rates of steel wire embedded in mortar cylinders [32] by observing that corrosion can be inhibited by mixing CMA and NaCl, only for concentrations of about 0.05 M to 0.1 M NaCl, while CMA admixture does not prevent corrosion when the concentration of NaCl is more than 0.2 M. Similar results were obtained for sodium acetate additions. Another study conducted in Oklahoma, US [21], indicated that although CMA solutions ponded on chloride-contaminated reinforced concrete slabs were not able to passivate the reinforcement, the same solutions did not cause any corrosion on reinforcement embedded in uncontaminated concrete slabs. This result is supported by a 4-year long study by Chollar, B. H. & Virmani, Y. P[24], which reported that after four years of ponding on reinforced concrete slabs with NaCl and CMA solutions, no corrosion was produced by the CMA solutions in contrast to the heavy corrosion and subsequent cracking of the concrete produced by the NaCl solutions. Similar conclusions were drawn by two studies that compared the electrochemical behaviour of metals immersed in solutions of various concentrations of CMA and NaCl [33, 34]. In general, it can be said that CMA solutions do not induce corrosion of metals in almost any condition and in some cases even act as inhibitors.

Effects of Acetate-Based De-Icers On Concrete

Several studies have been carried out over the past two decades focussed on the effects of CMA deicers on the properties of concrete in contact with the applied deicer. In some cases, contradictory results have been recorded for similar kinds of tests by different researchers. This is mainly due to the diversity of conditions under which those experiments were undertaken. The effect of CMA can act very differently when in contact with concrete depending on factors such as ambient temperature, Ca/Mg molar ratio of the CMA, concentration of the CMA solution, concrete mix design, cement composition, possible admixtures, type of aggregates etc. Consequently, the results from various studies cannot be directly compared. Nonetheless, they do offer an indication of the mechanism that governs the changes, which take place in concrete microstructure as well as in the composition of the CMA solution when these come in contact. Santagata & Collepardi [20] record severe deterioration of concrete after long-term, continuous immersion in concentrated CMA solutions (25% wt) at a temperature of 20oC. Another study by Lee et al [35]), involved tests at 58oC and found CMA to be particularly destructive due to the formation of expansive brucite and non-cementitious MSH. They also noted that the greater the Mg/Ca molar ratio, the greater the damage. This latter observation is also supported by a paper published in 1995 [18] which also documents that CMA solutions are more destructive in higher temperatures. However, this last study concluded that dilute solutions of CMA have a greater effect on compressive and flexural strength than concentrated solutions. The author went on to speculate that, this could be due to the fact that most of the magnesium in the solution is being consumed in the formation of $Mg(OH)_2$, and thus, most of the solution ends up comprising mostly calcium acetate, which he considers especially deleterious to concrete, although this is not supported by the tests involving this substance, conducted by Lee in his aforementioned study. However, most of the existing literature on the subject, support the concept of the formation of brucite and potentially the subsequent replacement of CSH in concrete with non-cementitious MSH as a result of the contact of concrete and CMA. Higher temperatures seem to have a detrimental effect on the cementitious matrix, and higher Mg/Ca molar ratios of the CMA appear to make it more destructive towards concrete. Higher concentrations of CMA solutions have a detrimental effect as documented in the literature, although some conflict exists on this matter, as one study has found dilute solutions to be more destructive.

Environmental Effects of Acetate De-Icers

Modern concepts of sustainability and whole-life costing of structures cannot be complete without consideration for the environmental effect on the construction materials or how the use of structure during the period in service affects the environment. For highway structures, some of the basic environmental considerations, in the case of de-icers applied during the winter months, is whether they can alter the physicochemical properties of soil; their effect on water and air quality or on vegetation. Several studies have been conducted to this end and they have found CMA and NAAC to be environmentally low-risk de-icers in comparison with the chloride-based ones. More specifically, CMA has been found to be beneficial to soil [36, 37, 38] and harmless to vegetation [39, 4]. Certain concerns have been expressed, mainly regarding the possible future effects of the acetate-based de-icers on surface waters and treatment ponds[14, 38]. However, practical application of both CMA and NAAC has indicated no such effect [29]. A summary of the conclusions of some notable studies carried out are presented in Table 4 below.

Table 4. Environmental effects of de-icers

DE-ICING AGENT	EFFECTS ON SOIL	EFFECTS ON VEGETATION	EFFECTS ON WATER QUALITY
NaCl	Potentially damaging. Exchangeable Sodium Percentage (ESP) more than 13 (toxic) [39].	Salt spray causes leaf burn. ESP higher than 4 kills some species. 0.5% concentration of chloride in plant tissue will result to plant death [39]. Browning, falling leaves, stem dieback, stunted or abnormal growth, premature death. Effect more apparent on grass rather than trees [14]..	Sodium and chloride concentrations in water supplies often higher than the standards set by health organisations [39].

CMA	Beneficial by increasing the pH and decreases solubility of trace metals. No upper limit defined [36, 37, 38]	No apparent adverse effect by spraying. Plants can withstand root exposure of up to 2500mg/l. [39, 4].	Biodegradable. No effects to water biota. It is better if the dilution rate is more than 100:1 [38]. It could cause eutrofication due to high BOD when reaeration of waterways is limited by ice cover or in poorly flushed ponds [14]. It could significantly increase organic loading in treatment works when used in urban environments [40]. Despite concerns about BOD and DO levels of receiving waters practical experience has recorded no adverse effects.[29].
NAAC	No reports	No reports	Despite concerns about BOD and DO levels of receiving waters practical experience has recorded no adverse effects [29].

CASE STUDY

Objectives

The main objectives of the laboratory study are:

- To monitor the corrosion rates, if any, of steel reinforcement in concrete exposed to CMA and Sodium Acetate solutions and to compare them with the corresponding effects on steel reinforcement in concrete exposed to NaCl solutions and water.
- To monitor the effects of varying the concentration of the solution of acetate-based de-icers in relation to their ability to prolong the time to corrosion initiation and to suppress corrosion rates.

- To investigate whether and how the addition of microsilica, varying the w/c ratio of concrete and the depth of cover can influence the time to corrosion initiation and corrosion rates if samples undergo the same testing regime.
- To determine whether the exposure of concrete to the de-icers under investigation could compromise the passivity of the rebars in the future, by causing carbonation of the concrete cover and thus alter the pH of the pore solution.
- To develop an understanding of the parameters controlling the behaviour of acetate-based de-icing agents in relation to the corrosion of steel reinforcement in concrete and their interaction with Portland cement and microsilica-containing paste by means of a microstructural investigation.

Methods

Corrosion monitoring

The specimens used for this particular part of the study, had to be manufactured in such a way as to cover the needs of the proposed investigation. Thus, small-scale slab specimens (210mm x 330mm 100mm) were manufactured with three sets of three reinforcing bars each. The three sets of bars were placed at three different depths from the bottom of the ponding basin. Specifically, the depths chosen were 10mm, 25mm and 40mm, as shown in Figure 2.1(a) and 2b. The purpose of employing different cover depths was i) to establish the time to corrosion initiation induced by the different solutions at different depths and ii) to investigate the effect of the cover depth on the corrosion rates of the reinforcement which may, in turn, reveal information regarding the interaction between the de-icer and the concrete cover.

Steel electrodes were made from 8mm diameter mild steel bars. The bars were initially cut to a 310mm length each, and a 4mm hole was drilled on one side at a distance of 10mm from the edge of the bars. They were then turned to a bright steel finish and degreased with acetone in order to remove any manufacturing grease. Next, protective coating was applied on both ends of each bar, which served the purpose of ensuring that an equal area of steel was exposed in every bar, a detail essential as the anodic area has to be known for the calculation of the corrosion current. It also ensures that the protruding ends of the bars are protected from aggressive ions and are, therefore, not corroding. The protective coating was applied by dipping 95mm of either end of the bar thus, ensuring an exposed length of 120mm in the middle. The coating comprised initially two coats of binder paste with the same binder and the same water/binder ratio as the specimen in which the bars were to be imbedded in each case. Two coats of marine varnish were applied, also by dipping, to complete the coating process. The bars were then placed in the moulds as described above and the depth of cover checked.

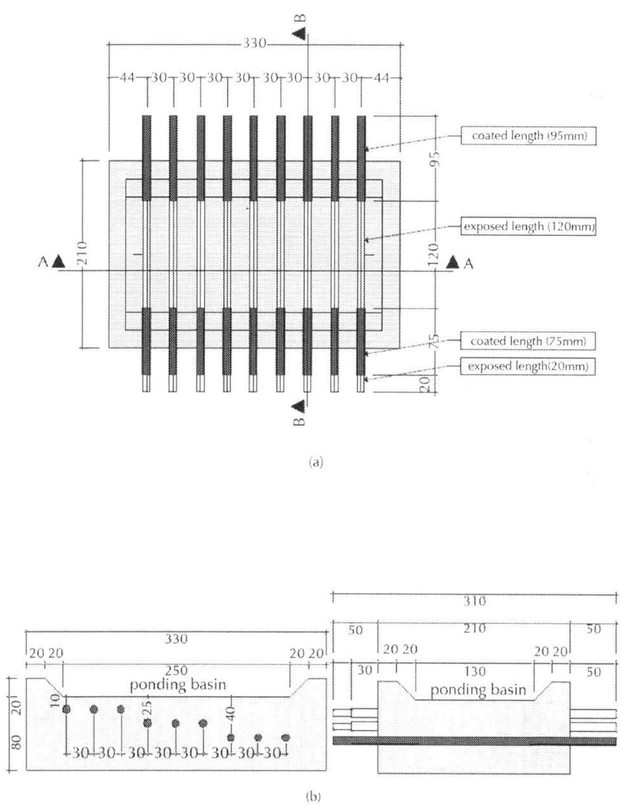

Figure 2: Constructional details of corrosion monitoring specimens, (a) Plan view, (b) Section A/A (c) Section B/B.

Standard batching and mixing techniques were adopted, and the same procedure was adopted for each mix. The mixer used for the preparation of all mixes was a Zyklos ZZ30. When filled and vibrated, the moulds were placed in the curing room (20 oC, 100% R.H.) where they were left to hydrate for 24 hours. Following that period, they were de-moulded and returned to the curing room for a further 27 days before testing. On the 27th and 28th days of curing two coats of anti-slip floor coating were applied on all sides of the specimens, including the protruding ends of the steel bars, except the ponding basin. After curing, the specimens were transferred to a room with constant temperature and relative humidity (20 oC and 70% R.H.), where they remained for the rest of the experimental process. Immediately after their removal from the curing room they were ponded with 300g of water saturated with calcium hydroxide. Fortyeight hours later the initial base-line measurements were taken as described below. Thereafter, two-weekly dry-wet cycles (seven days dry, seven days wet) were followed whereby, the wet

period of each cycle involved ponding with 300g of the corresponding solution for each

specimen and sealing to prevent evaporation. Measurements were taken at the end of every wet half-cycle in order to ensure conductivity between the surface of the ponding basin and the steel bars.

Table 5. Slab specimens for corrosion measurements

TYPE OF BINDER	W/B RATIO	NUMBER OF SLAB SAMPLES	NUMBER OF REINFORCEMENT BARS
OPC	0,40	6	54
	0,65	6	54
90%OPC / 10% Silica fume	0,40	6	54
	0,65	6	54
Total		24	216

It becomes obvious, therefore, that six samples were manufactured for each one of the four mixes designed for the project. This was done for comparison purposes as there were six different aqueous solutions chosen for ponding on the samples. The chosen solutions were as follows: i) Water, ii) 3.5% wt. NaCl solution, arbitrarily chosen to correspond to the average concentration of NaCl in sea water, ii)3.5% wt. CMA solution, chosen due to the indications from practical experience in the literature that equal amounts of CMA and NaCl produce equal de-icing performance, iv)6.181% wt. CMA solution, chosen to produce the same freezing point as 3.5% wt. NaCl, v)3.5% wt. NAAC solution, chosen due to the indications from practical experience in the literature that equal amounts of NAAC and NaCl produce equal de-icing performance, and vi)4.367% wt. NAAC solution, chosen to produce the same freezing point as 3.5% wt. NaCl.

Two well established types of corrosion measurement techniques were used so that their combined results could provide an insight into whether corrosion is taking place and what the corrosion rate is at any given time. The two techniques used were the half-cell potential measurement and Linear Polarisation Resistance technique. The corrosion measurements were conducted by the use of a Sycopel Scientific Superstat potentiostat that was operated manually throughout the duration of the experiments for most of the samples with the exception of eight samples. Namely, the two samples with pure OPC binder and a w/c ratio of 0.65 that were ponded with CMA solutions and all the six samples with a w/c ratio of 0.65 and a 10% silica fume admixture were tested using the Gamry PCI4-750 potentiostat after the completion of the 19th ponding cycle. A saturated KCl calomel electrode was used as a reference electrode. In both techniques, three measurements were taken at every depth for each sample. This provided an indication of the consistency of the results obtained.

pH shift measurements

A pH meter was used to measure the pH of the 300g of every ponding solution right before ponding and then right before the solutions were removed at the end of the wet cycles throughout the duration of the corrosion monitoring experiments. The negative and positive pH shifts were recorded and plotted over time. This was done to provide some means of observing the interaction between the solution and the specimen and to offer an indication of the exchange of ionic species and compounds between them.

Scanning Electron Microscope Imaging

Small cement paste prisms (10 mm x 10 mm x 110 mm), were cast with a water/cement ratio of 0.4 and were immediately placed in the curing room at 100% RH. The following day they were de-moulded and placed back in the curing room for a further period of 27 days. After curing, the samples were transferred to a room maintained at constant temperature and relative humidity (20°C and 70% RH), where they remained for the rest of the experimental process. They were then coated with resin and left to dry overnight. The coating process was repeated the following day and they were, again, left to dry overnight. The following day the hardened resin along with the as-cast surface was removed from one of the two smaller faces of the sample, with a diamond saw. This way, it was ensured that the penetration of the deicer solution from that contact face would be unidirectional so that the depth of penetration could be estimated with higher precision. Moreover, the removal of the as-cast surface of that particular face ensured that misinterpretations due to contamination would be avoided. The immersion regime, which the samples underwent, involved two-weekly cycles (one week immersed in solution and one week dry). Although wet-dry cycles may bear no particular significance in this case, the experiment was designed to correlate with the ponding regime performed on miniature slab samples designed for reinforcement corrosion testing, which forms the main core of this general project.

The effect of the immersion process was microscopically examined at predetermined times for a duration of 1 year after the end of the curing process, namely at 28 days, 90 days, 6 months and 1 year. The examination was performed with the help of a HITACHI 3200 scanning electron microscope. Specifically, the samples were cut in the middle with a diamond saw along their longitudinal axis, which, contrary to thin slicing, allows for visualisation of the effects of the solutions on the same sample, simply by moving the cut profile under the microscope. Samples were carbon coated before taken to the SEM for imaging. Images were taken at various locations, at different depths from the surface of the exposed face and at different magnifications, which varied from x300 to x9000. Element mapping was performed with the help of the microscope in order to help identify the elements and compound formations at locations of interest.

RESULTS AND DISCUSSION

In what follows, only selected parts of the results obtained from the different sets of experiments, which forms the key part of the data are presented.

Corrosion monitoring

The "Van Daveer" criteria for the interpretation of half-cell potential measurements taken with a copper/copper sulphate reference electrode specify that a potential more positive than -200mV indicate less than 10% probability of corrosion and a potential more negative than -350mV indicates more than 90% probability of corrosion [41]. The electrode used for the halfcell potential measurements was a saturated calomel electrode whose potential is 72mV more positive than the potential of a copper/copper sulphate electrode. Thus, a half-cell potential more negative than -278mV, indicates a 90% probability of corrosion. Similarly, potentials more positive than -128mV represent an indication of a less than 10% probability of corrosion. Values in between represent an uncertain corrosion activity [42]. Figures 3. and 4., illustrate the results produced so far by the measurement of the rest potential of the steel bars in OPC and 90% OPC / 10% Silica Fume specimens with a water/cement ratio of 0.65.

The results presented in figures 3. and 4.; concern only the half-cell potential measurements taken from the OPC and OPC/Silica Fume samples respectively, with a w/c ratio of 0.65. Clearly, no corrosion seems to be occurring in any of the corresponding samples with a w/c ratio of 0.4 after 25 wet-dry cycles.

In the OPC samples the only bars which seem to be exhibiting signs of a high probability of corrosion are the bars situated at 10mm depth in the OPC sample ponded with a 3.5% aqueous solution of NaCl (figure 3.a and 3.b). The bars at 25mm depth in the same specimen show a tendency to corrode but the half-cell potential values obtained are more positive than the value considered characteristic of a 90% probability of corrosion according to the aforementioned criteria. At the end of the 25th cycle these middle bars have been measured to have a potential of -187mV which is a sign of uncertain corrosion activity. The half-cell potential values of the bars at 40mm depth dwell at around -100mV, which is a value characteristic of a very low probability of corrosion. The difference in potential between bars situated at these three different depths probably results from differences in the concentration of free chloride ions around each bar. The free chloride ions can disturb the passive film on the surface of the steel bars an therefore act as catalysts towards the initiation of corrosion. In addition, the dissociation of iron chlorides can produce locally high levels of acidity at the surface of the steel leading to pit formation (Pullar-Strecker 2002). A careful observation of Figure 3. shows that, with the exception of the sample ponded with the NaCl solution, each one of the rest of the samples exhibits comparable values of rest potential over time at all three depths. The significance of this observation lies in the fact that it indicates the absence of diffusing aggressive ions, which could, directly cause corrosion after reaching a certain threshold, in the same way that chloride ions do. However, the absence of chlorides in a solution does not render it incapable of inducing rebar corrosion as its contact with concrete over a longer period of time could have other undesirable effects such as lowering of the pH of the concrete cover which would, in turn, render the reinforcement susceptible to corrosion.

It is also obvious in Figures 3.a and 3.b that, although the acetate solutions and the water do not seem to corrode the reinforcing bars after 25 ponding cycles, each one of those solutions produces a distinct pattern of potential values over time. The trend evident in the case of the sample ponded with the CMA solution closely resembles the trend produced by the potential of the bars in the sample ponded with water. This could mean that CMA does not contribute any ions more than water does, to the pore solution. Otherwise, this behaviour could indicate that CMA produces impervious deposits on the surface of the concrete, which block the capillary system of the cement paste and hinder the ingress of ions in the pore solution. In contrast, sodium acetate, at all three depths seems to produce the same behaviour as the graph that corresponds to the potentials measured at the lower bars, located at the 40mm depth, in the samples ponded with the NaCl solution. The chloride ions around these bars have not reached a threshold concentration, which could initiate depassivation and corrosion after 25 cycles of testing. In both cases of the samples ponded with NaCl and the ones ponded with NAAC, the pore solution has been influenced as indicated by the lower rest potentials of these bars compared to the rest potentials of the samples ponded with water. This similarity in patterns could be due to the ingress of sodium ions that are present both in NaCl and in NAAC, on the pore solution of the concrete.

The same patterns are evident in Figure 4., which illustrates the corresponding results for the samples containing Silica Fume. However, in these samples none of the bars has had the time to corrode within the 25 cycles of testing and the only bars which produce lower rest potentials and probably a tendency to initiate corrosion are the bars situated at a depth of 10mm from the surface in the sample ponded with NaCl solution. This is due to the documented effect of silica fume to alter the pore size distribution of the cement paste, reducing the size of the capillary pores, thus, making the concrete more impermeable and more resistant to attacks by chlorides and other aggressive ions [40]. However, in all the samples containing silica fume that were ponded with NAAC solutions the rest potential values tend to dwell consistently around 100 mV more positive than they do in the OPC samples. This could mean that the effects of NAAC on the pore solution are more evident in the samples with larger capillary pores, namely the OPC samples.

Finally, it should be noted that all samples were sectioned and sprayed with a phenolphthalein and alizarin pH colour indicators which showed no sign of reduced pH near the surface and therefore no carbonation taking place in any of them. The bars were extracted from all samples and were visually inspected; they all supported the results of the non-destructive corrosion monitoring, as corrosion was encountered only at the bars closest to the surface in the OPC samples with a w/c ratio of 0.65.

ph Shift

Figures 5. and 6. illustrate the results for the pH shift monitoring described in section 3.2 above throughout the duration of the corrosion monitoring experiments. Although such a technique does not provide solid proof of any microstructural process, the results can be used as supporting evidence for the data derived from the widely used microstructural investigation techniques employed throughout this project. A simple observation of the information on figures 5. and 6. reveals a number of things worth noting.

The general trend observed is that all the plots tend to gradually decrease over time from more positive to more negative values and to eventually stabilise or at least decrease at a lower rate. This is possibly due to leaching of $Ca(OH)_2$ from the concrete and the subsequent dissociationof the calcium and hydroxyl ions, with the latter raising the pH value of the solution. The calcium hydroxide, however, becomes less available as the hydration process progresses and consumes it and the pores diminish in size. It also becomes less available due to the fact that much of it, previously located on and very close to the surface of the ponding basin of the samples has leached out during the initial stages of ponding. This latter phenomenon is probably what causes the abrupt drop of the pH shift graphs towards less positive (or more negative) values, within the first two to four cycles in most samples. The general trend observed is that all the plots tend to gradually decrease over time from more positive to more negative values and to eventually stabilise or at least decrease at a lower rate. This is possibly due to leaching of Ca(OH)2 from the concrete and the subsequent dissociation of the calcium and hydroxyl ions, with the latter raising the pH value of the solution. The calcium hydroxide, however, becomes less available as the hydration process progresses and consumes it and the pores diminish in size. It also becomes less available due to the fact that much of it, previously located on and very close to the surface of the ponding basin of the samples has leached out during the initial stages of ponding of the calcium and hydroxyl ions, with the latter raising the pH value of the solution. The calcium hydroxide, however, becomes less available as the hydration process progresses and consumes it and the pores diminish in size. It also becomes less available due to the fact that much of it, previously located on and very close to the surface of the ponding basin of the samples has leached out during the initial stages of ponding. This latter phenomenon is probably what causes the abrupt drop of the pH shift graphs towards less positive (or more negative) values, within the first two to four cycles in most samples. The general trend observed is that all the plots tend to gradually decrease over time from more positive to more negative values and to eventually stabilise or at least decrease at a lower rate. This is possibly due to leaching of $Ca(OH)_2$ from the concrete and the subsequent dissociation of the calcium and hydroxyl ions, with the latter raising the pH value of the solution. The calcium hydroxide, however, becomes less available as the hydration process progresses and consumes it and the pores diminish in size. It also becomes less available due to the fact that much of it, previously located on and very close to the surface of the ponding basin of the samples has leached out during the initial stages of ponding.

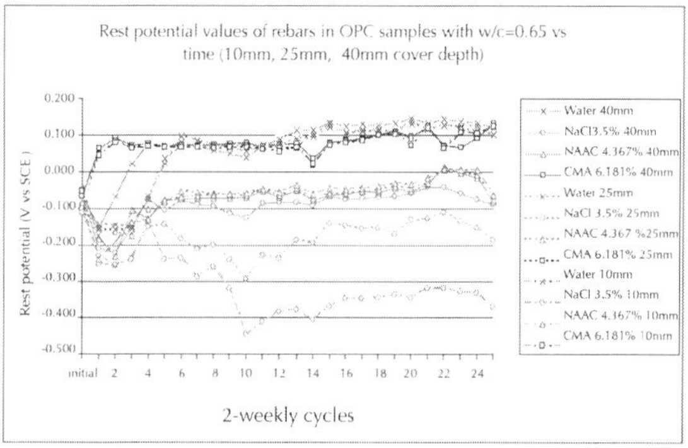

(a)

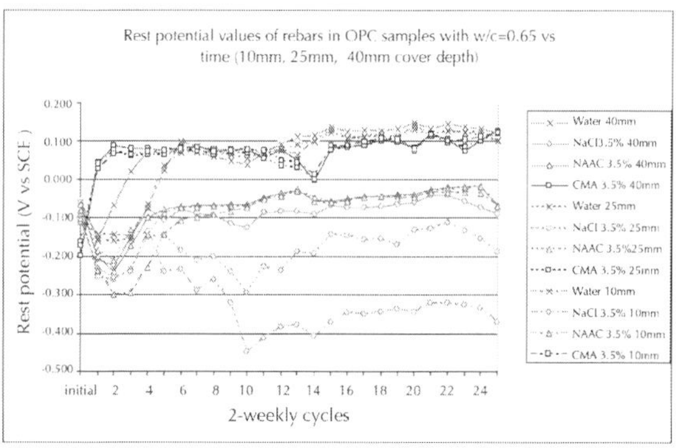

(b)

Figure 3: Half-cell potential values of rebars in OPC samples w/c=0.65 vs. time (10mm, 25mm, 40mm cover depth), a) Samples ponded with solutions with a freezing point equal to that of a 3.5% wt NaCl, b) Samples ponded with solutions of 3.5% wt.

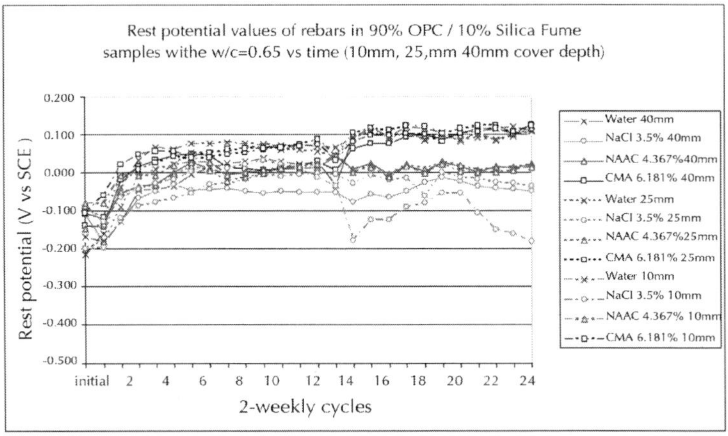

(a)

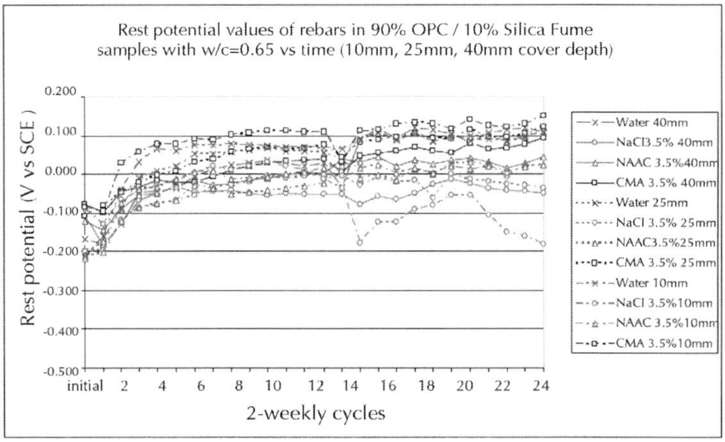

(b)

Figure 4: Half-cell potential values of rebars in 90% OPC/10% Silica Fume samples w/c=0.65 vs. time (10mm, 25mm, 40mm cover depth), a) Samples ponded with solutions with a freezing point equal to that of a 3.5% wt NaCl, b) Samples ponded with solutions of 3.5% wt.

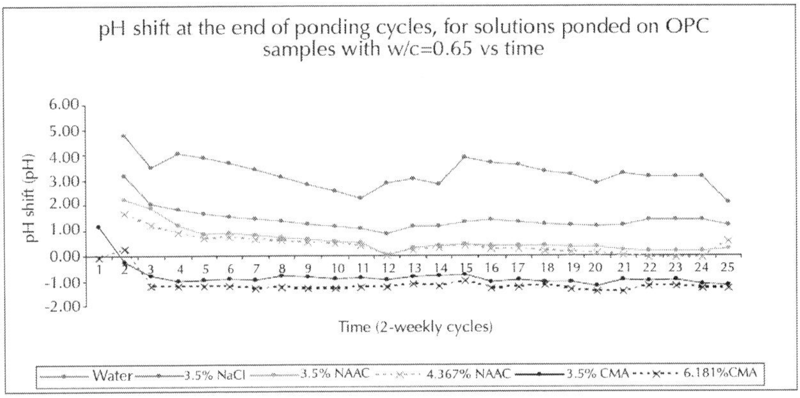

(a)

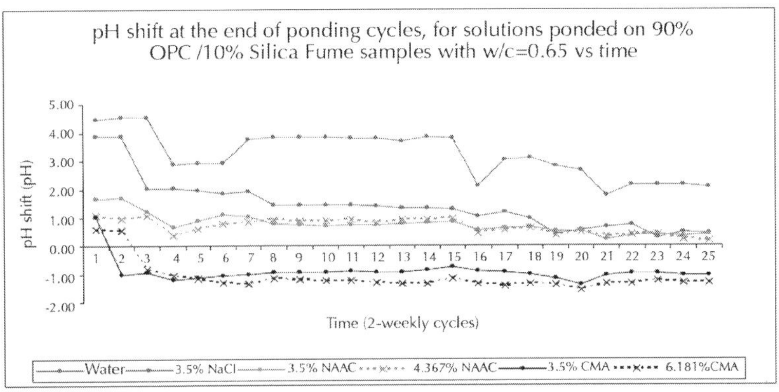

(b)

Figure 5: pH shift in ponding solutions at the end of ponding cycles for samples for samples with w/c=0.65, (a) OPC samples, (b)90% OPC / 10% Silica Fume samples.

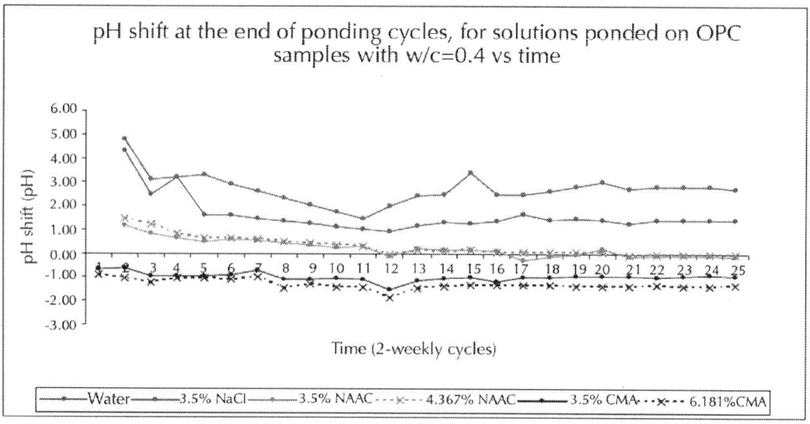

(a)

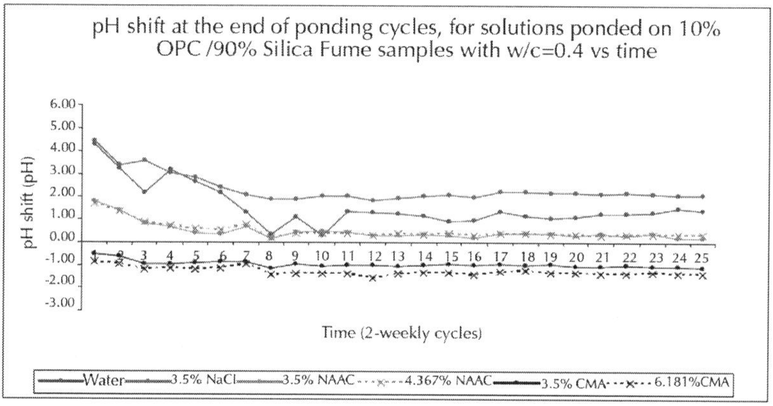

(b)

Figure 6: pH shift in ponding solutions at the end of ponding cycles for samples for samples with w/c=0.4, (a) OPC samples, (b)90% OPC / 10% Silica Fume samples.

Varying the water cement ratio does not seem to have a major effect on the pH shift, as one might expect. This could be partly due to the fact that the samples with a 0.65 w/c ratio have larger pores while in the 0.4 w/c ratio samples, more calcium hydroxide is available. Therefore, leaching of calcium hydroxide could be facilitated in these two different ways, thus, producing seemingly comparable results.

One, however, has to note that in the case of the CMA solutions on the 0.65 w/c ratio samples in figure 5., the graph decreases more abruptly producing positive pH shift initially within the first few cycles, than it does in the case of the corresponding solutions on samples with 0.4 w/c ratio as illustrated in figure 6. where the shift is negative from the onset of the measurements.

The 10% replacement by silica fume does not seem to have a significant effect on the pH shift for any of the solutions. Moreover, the variations in concentration of the acetate-based de-icer solutions seem to play no significant role either. This is possibly accounted for by the fact that the variations in concentration of the acetate de-icer solutions chosen for this project are relatively small.

It is interesting to note that the resulting pH shift for the CMA solutions is predominantly negative while for the NAAC solutions it is positive or in some cases marginally positive, throughout the graphs. Thus, despite the fact that NAAC and CMA solutions are both acetate solutions, equal or comparable concentrations of the two in this experiment, consistently produce a difference in pH shift at the end of ponding. This effect could be due to the strong presence of magnesium in the CMA. It is worth noting that the Ca/Mg molar ratio in the CMA used is 3/7. Therefore, when soluble lime leaches from the concrete into the CMA solution, and calcium and hydroxyl ions are released, dissociated magnesium is bound to form magnesium hydroxide otherwise known as brucite, a very insoluble and relatively impermeable compound. This possibly led to the decreased values of the pH at the end of the ponding cycles in CMA solutions.

CONCLUSIONS

This case study suggests that:

1. Acetate based de-icers, unlike NaCl, do not contain deleterious ions which cause the corrosion of the steel reinforcement bars.
2. Presence of Acetate de-icers neutralised the probability of corrosion.
3. The trend for corrosion probability evident in the case of the sample ponded with the CMA solution closely resembles the trend produced by the bars in the sample ponded with water.

This could mean that CMA does not contribute any ions more than water does, to the pore solution.

4. NAAC and CMA solutions of equal or comparable concentrations, consistently produce a difference in pH shift at the end of ponding.
5. Varying the water cement ratio does not seem to have a significant effect on the pH shift, as one might expect but no corrosion seems to be occurring in any of the samples with a w/c ratio of 0.4 even after 25 wet-dry cycles.
6. The use of Acetate-based de-icers on reinforced concrete would lead to low maintenance and repair cost for concrete bridges.

RECOMMENDATION

For better **corrosion** prevention in highway structures, the following preventive strategies need to be considered:

1. The high cost of Acetate based de-icers could be solved by making a change in the deicing tradition and protocols and on highways..
2. Reinforced concrete bridges may need to locally de-iced with the more expensive but noncorrosive Acetate-based de-icers.
3. There is need to increase the general awareness of the very high corrosion costs and potential savings.
4. There is need to change the misconception that nothing can be done about corrosion rather efforts should not be spared in the quest to reduce cost by effecting adequate change in existing policies, regulations, standards, and management practices to increase corrosion savings through sound corrosion management.

REFERENCES

1. TRB, Highway Deicing, Comparing salt and calcium magnesium acetate Transportation Research Board, National Research Council, Washington D.C., Special Report 235, 1991.
2. Pullar-Strecker, P., Concrete Reinforcement Corrosion, from assessment to repai decisions, 1st edn, Thomas Telford Ltd, London, 2002.

3. Richardson, M. G., Fundamentals of Durable Reinforced Concrete, 1st edn, F & EN Spon Press, London, 2002.
4. Robidoux, P. Y. & Delisle, C. E., "Ecotoxicological evaluation of three deicers (NaCl, NaFo, CMA) - Effect on terrestrial organisms", Ecotoxicology and Environmental Safety, vol. 48, no. 2, pp. 128-139, 2001.
5. Slick, D. S., Effects of calcium magnesium acetate on pavements and motor vehicles, 1988.
6. G.H. Koch, M.P.H. Brongers, N.G. Thompson, Y.P. Virmani, and J.H. Payer, Corrosion Costs and Preventive Strategies in the United States," U.S. Federal Highwa Administration Report, FHWA-RD-01-156, March 2002.
7. McCrum, R. L., Corrosion Evaluation of Calcium Magnesium Acetate (CMA), Salt (NaCl), and CMA/Salt Solutions, Michigan Department of Transportation, Research Laboratory Report R-1255, 1988.
8. McCrum, R. L., "Calcium Magnesium Acetate and Sodium-Chloride As Highway Deicing Salts - A Comparative-Study", Materials Performance, vol. 28, no. 12, pp. 24-28, 1989.
9. Louis D.Johnston & Samuel H.Williamson. The Annual Real and Nominal GDP for the United States, 1790 - Present.. Economic History Services. 2005.
10. JACC, Survey of Corrosion Cost in Japan, Japan Association of Corrosion Control, 1999.
11. Dori, J. & Fisher, R. D. Jr., U.S And Asian Statistical Handbook 1999-2000 Th Heritage Foundation, Washington, 2007.
12. Wallbank, E. J., The Performance of Concrete in Bridges, A survey of 200 Highway Bridges, Her Majesty's stationery office, London, 1989.
13. CRC Press, Handbook of chemistry and physics : a ready-reference book of chemical and physical data, 81st edn, Boca Raton, Fla., London, 2000.
14. Mangold, T., Road Salt Use for Winter Maintenance.A Review of Impacts, Alternatives, and Recommendations for the St. Paul Campus Stormwater Management Plan, 2004.

15. Mauritis, M., Mcgraw, J., & Jang, J. W., "Method for Determining Ice Undercut Temperature of Deicing Chemicals", Journal of Materials in Civil Engineering, vol. 7, no. 1, pp. 84-86, 1995.
16. Bensted, J., "Chemical effects of cement mortar of calcium magnesium acetate as a deicing salt - Discussion", Cement and Concrete Research, vol. 26, no. 4, pp. 633-635, 1996.
17. Fujii T. "Combined effect of acid and deicer solutions on fiber-reinforced hardened cement pastes", in 2nd International Conference on Concrete under Severe Conditions, Sakai K., Banthia N., & Gjorv O.E., eds., E & FN Spon.
18. Peterson, O., "Chemical Effects on Cement Mortar of Calcium-Magnesium Acetate As A Deicing Salt", Cement and Concrete Research, vol. 25, no. 3, pp. 617-626, 1995.
19. Peterson, O., "Chemical effects of cement mortar of calcium magnesium acetate as a deicing salt - Reply", Cement and Concrete Research, vol. 26, no. 4, pp. 637-639.
20. Santagata, M. C. & Collepardi, M. 2000, "The effect of CMA deicers on concret properties", Cement and Concrete Research, vol. 30, no. 9, pp. 1389-1394, 1996.
21. Locke, C. E., Kennelley, K. J., Boren, M. D., & Luster, V., Study of corrosion properties of a new deicer, Calcium Magnesium Acetate, 1987.
22. Fischel, M., Evaluation of selected deicers based on a review of the literature Colorado Department Of Transportation, CDOT-DTD-R-2001-15, 2001.
23. Katsanos A. & Nwaubani S.O. "Viability of Alternative de-icing operations on European highway bridges", Global Construction: Ultimate Concrete Opportunities, ed..
24. Chollar, B. H. & Virmani, Y. P., "Effects of calcium magnesium acetate on reinforced concrete steel", Public Roads, vol. 51, no. 4, pp. 113-115, 1988.
25. Harrach, N. & Wyatt, J., "Fine tuning CMA for corrosion control", Public Works, vol. 121, no. 8, pp. 40-41, 1990.
26. [Harris, G. P., Turner, R., & Nelson, R. J., "Field-Test Comparison of Calciu Magnesium Acetate to Salt", Journal of Transportation Engineering-Asce, vol. 119, no. 6 pp. 889-904, 1993

27. Miller, W. L. CMA Cuts Corrosion on the Zilwaukee Bridge. Better Roads [63], 1 19. 1-2-1993.
28. Miller, W. L., "CMA Use on the Zilwaukee Bridge", Chemical Deicers and the Environment pp. 553-557, 1992.
29. The City of Aspen. Water Quality. www.aspenpitkin.com/depts/44/waterquality.cfm. 2002b.
30. The City of Aspen. Air Readings. www.aspenpitkin.com/depts/44/air_readings.cfm. 2002a. Ref Type: Electronic Citation
31. Schenk, R., Ice Melting Characteristics of Calcium Magnesium Acetate, FHWA Washington D.C., FHWA/RD-86/005, 1985.
32. Ushirode, W. M., Hinatsu, J. T., & Foulkes, F. R., "Voltammetric Behavior of Iron in Cement, Effect of Acetate and Urea Additions", Journal of Applied Electrochemistry, vol. 22, no. 3, pp. 224-229, 1992.
33. Kennelley, K. J. & Locke, C. E., "Electrochemical-Behavior of Steel in Calciu Magnesium Acetate", Corrosion, vol. 46, no. 11, pp. 888-895, 1990.
34. Slick, D. S., Effects of calcium magnesium acetate on pavements and motor vehicles 1988.
35. Lee, Hyomin, Robert D. Cody, Anita M. Cody, and Paul G. Spry. "Effects of various deicing chemicals on pavement concrete deterioration." In Center for Transportation Research and Education (Ed.), Mid-Continent Transportation Symposium Proceedings, Iowa State University, Ames, USA, pp. 151-155. 2000.
36. Amrhein, C., Mosher, P. A., Strong, J. E., & Pacheco, P. G., "Heavy-Metals in th Environment - Trace-Metal Solubility in Soils and Waters Receiving Deicing Salts", Journal of Environmental Quality, vol. 23, no. 2, pp. 219-227, 1994.
37. Amrhein, C., Strong, J. E., & Mosher, P. A., "Effect of Deicing Salts on Metal an Organic- Matter Mobilization in Roadside Soils", Environmental Science & Technology, vol. 26, no. 4, pp. 703-709, 1992.
38. Horner, R. R. & Brenner, M. V., "Environmental Evaluation of Calcium Magnesium Acetate for Highway Deicing Applications", Resources Conservation and Recycling, vol. 7, no. 1-3, pp. 213-237, 1992.

39. Moran, V. M., Abron, L. A., & Weinberger, L. W., "A Comparison of Conventional and Alternative Deicers - An Environmental-Impact Perspective", Chemical Deicers and the Environment pp. 341-361, 1992.
40. Rabideau Alan J.. & Matsumoto Mark R. 1987, "Impact of Calcium Magnesiu Acetate Road Deicer on POWT Operation", Journal of Water Resources Planning and Management, vol. 113, no. 2, pp. 311-315.
41. Pullar-Strecker, P., Concrete Reinforcement Corrosion, from assessment to repai decisions, 1st edn, Thomas Telford Ltd, London, 2002.
42. Bjegovic, D., Mikulic D, & Sekulic D, "Non-destructive corrosion rate monitoring for reinforced concrete structures", in 15th World Conference on Non-Destructive Testing, WCNDT, 200.

Chapter 4

Street Furniture and Amenities: Designing the User-Oriented Urban Landscape

Gökçen Firdevs Yücel

Faculty of Engineering and Architecture, Istanbul Aydın University, Istanbul, Turkey

INTRODUCTION

It is often difficult to understand fully how site design integrates the elements of furniture that are used for various designated purposes and create a sense of place. Such an understanding can open up creative possibilities for designers of outdoor areas and enable them to shape spaces that can succeed in achieving their aims over the long term [1].

If street furniture is properly integrated in the design of a public space, it creates an identity and develops a sense of place around it [2]. Items of street furniture comprising, for example, objects used to facilitate transportation or the use of land adjacent to a street, are commonly used in urban areas with the aim of making the street more aesthetically appealing; however, these objects, if placed near intersections, may obstruct the view of drivers, especially if crowds develop around them; and if they are close to the roadway, they should also not obstruct the sight distance of pedestrians [3].

The furniture should be selected and set up based on an analysis of the site's current and desired patterns of use, so it can serve its purpose effectively; quality furniture is costly, so it should be used only where it is really needed. There are various approaches to selecting or designing street furniture: for example, there could be a coordinated selection which gives a consistent tone to streets and walkways; or various parts of the streetscape could be designed to serve as artworks in themselves [4].

Generally speaking, there are five basic criteria involved in selecting and placing items of street furniture: function (i.e. seeing how necessary an item is and how it can serve its purpose), siting and layout (i.e. deciding where each item should be places), form and appearance (i.e. making sure there a continuity or at least a linkage between the designs of different items), durability (given expected usage), and cost [5].

The Significance of Street Furniture

Street furnishings create the settings for resting, sitting and eating, and social encounters with others. Such settings may be of great importance to the elderly, those with limited mobility, and adults who have small children; but in addition to their functional aspect, items of urban furniture such as benches and tables in parks and squares can also be socially significant [6], as they give these sites a comforting and appealing air and draw people in together.

Appropriately selected and placed furniture can draw people to outdoor spaces and add to the pleasure of using these spaces; the main challenge is getting them outdoors, with the aim of making them feel welcome, relaxed, and involved [1]. The quality of urban spaces is indicated by their character and how well they create a sense of identity, as well as by the quality and placement of their street furniture, as can be seen in various cities: London's red telephone booths and Paris' metro entrances, for example, have become iconic and essential to these cities' identities [7]. In addition to their functional and symbolic roles, items of street furniture may also set standards and expectations of quality for the development of the areas where they are found [8] and can serve as talking points for planning [9].

FURNITURE PLANNING

In order to plan the design of street furniture, one must first plan establish and define the type of space under consideration – whether it is a park, a street, a plaza, a waterfront, a recreational area, or a mall, for example. The initial defining questions are: who are the potential users of the space, for what purpose and at what times will it be used, and how will it serve users when they arrive there? [1].

Typically, installations where the furnishings have been planned and coordinated as part of a broader design concept are more successful than those where they have been selected piecemeal without taking account of users' needs, or the architectural character or weather conditions at the site.

As regards what furniture should be consistent across various sites and which unique to a particular site, this depends on the situation, but in general there can be some consistent elements, such as signage, or streetlighting (for quality of illumination, aesthetics, and to facilitate maintenance by standardizing the types of bulbs used) [10]. In brief, the goal is to create convenient, easily maintained and publicly accessible amenities that do not hinder pedestrians or traffic [11].

Placement of Street Furniture

The location of the furnishings should be based on their functions [4], and coherent with the patterns and designs of the hard surfaces at the site. The items of furniture should not give the appearance of being cluttered, so that, for example, signs are attached to a single post or column rather than being distributed in several places; group amenities such as seating areas or telephone booths are set up where they will be used, such as at pedestrian junctions; and features like bollards are used minimally [12].

In addition, the furniture should not be hazardous for pedestrians; if it is used along a walking area it should be aligned to make passage easier, rather than staggered like an obstacle course [13, 14]. Varying the texture of the walking surface at key points can also help the visually impaired to find public amenities more easily [11].

STREET FURNITURE DESIGN

The furnishings of urban outdoor spaces should be designed with the needs of the disabled, children, and the elderly in mind, and also to enable triangulation – i.e. they should be linked together to stimulate social encounters and encourage people to talk to strangers.

In public spaces this stimulus can be created by the selection and arrangement of the street furniture; for instance, if benches, rubbish bins and telephones are far apart, they may have the effect of drawing people apart, whereas if they are arranged closely linked with other amenities such as a coffee cart, they tend to bring people together quite naturally.

Safety and Security

Furniture items designed for outdoor spaces must be constructed of safe materials and designed to prevent injury, without sharp edges or exposed fasteners. They are usually either attached to the ground with anchor bolts (for example, using surface mounting, i.e. attaching a bench to a concrete slab), or embedded in the ground. Naturally, the technique used for mounting should be decided on in advance, so the furniture can be made accordingly [15].

The type of furniture and its arrangement should also take into account visibility and sightlines, lighting, and accessibility issues that may be faced by women, children, the elderly and the disabled. Areas which are unsafe or pose risks should be designed taking these factors into consideration [16].

Materials Used in Street Furniture

Furniture selection and design should take into account weather effects such as sunlight, expansion and contraction, wind stress, moisture, and in some cases, salt spray, frost, or ice. The best designs usually incorporate strong, simple shapes, native materials, and natural finishes, generally in black, grays, and earth tones, accented with bright colors [17].

The most popular materials used are steel and wood; other possibilities are stone, concrete, recycled plastic and various other materials. The choice of materials depends on the context and limitations of the design; for example, whether the furniture should be resistant to vandalism, whether ventilation is needed for drying it during wet weather spells, what the weather conditions may be, how frequently the furniture is likely to be used and by whom, what the initial costs are, including mounting, the costs and ease of maintenance, and whether there is a possibility of using eco-friendly materials [15]. Site furniture should be made of nonflammable materials such as metal, brick, or stone, and wooden structures should be avoided in areas where there is a risk of fire [18].

Steel

Stainless steel is a popular choice of material, because it is versatile, has a high strength/weight ratio and can be shaped to fit almost any design aesthetic. It is durable, resistant to corrosion and impact, and requires no maintenance; it is also cheap compared to all the other materials in terms of life-cycle cost, and is fully recyclable.

Wood

Wood is another popular choice of material for street furniture, because it is a natural material that feels warmer in cold weather and cooler in hot weather, unlike metals. Its disadvantages are that it dries more slowly than metals after rain, needs more maintenance, and is easier to burn, break, carve graffiti into, or otherwise vandalize. Wood may be inexpensive, but the type of wood selected should depend on the location and frequency of use of the furniture. For example Pinus spp. is inexpensive and quite durable, but loses its color and longevity if it is not treated and maintained; the cost of maintaining or replacing it may outweigh the initial cost of purchasing it. Cedrus spp. weathers naturally, is non-toxic and has a pleasant odour, but it is not a hardwood, and can easily be carved into or chipped; it is therefore unsuitable for areas where there is frequent use of the furniture.

Concrete

Concrete furniture may not need to be mounted, as it is heavy; it is also difficult to vandalize (except with spray paint). Concrete has a long durability and may be aesthetically attractive if it is designed to be consistent with neighboring architecture; however, it does not drain or ventilate well after rain. Much of the concrete street furniture that exists today seems outmoded and appears to lack innovation.

Stone

Like concrete, stone is very durable, does not need to be mounted because it is heavy, and is resistant to vandalism except when spray paint is used. However, it has many of the same disadvantages: it does not ventilate or dry well after rain, and is limited in its aesthetic possibilities where design is concerned [15].

Plastic

Recycled furniture is generally made from plastics, which give it a smooth texture and appearance. The resulting furniture is highly durable, often lasting considerably longer than wood; low-maintenance; and therefore more cost-effective in the long term.

Colour

Items of street furniture should contrast significantly in color with the background where they are situated, and have a luminance contrast of at least 0.3 (30%) to increase their visibility to pedestrians. This can be supplemented by colored borders in the pavement, which mark the edges of the street furniture and guide pedestrians around it to clear and unobstructed walking areas [19].

Sustainability

Paint or other finishing materials of furniture should be non-toxic and non-staining [14]. The use of recycled materials in street furniture enables manufacturers to conserve natural resources and reduce their carbon footprints; it also educates the users of the furniture, the public, on the importance and mental and physical benefits of recycling. In addition, such furniture items can be recycled again when they need to be replaced, further reducing damage to the environment.

Local city administrations have been increasing initiatives to ensure the sustainability of their street furniture, including raising civic awareness, in order to reduce costs and fulfil legal requirements for emissions and energy efficiency. Most of these initiatives are demonstrably effective; replacing filament light bulbs with new LED bulbs, for example, can cut the energy consumption of street lighting by 40-70%.

With new, renewable energy sources other possibilities are emerging, such as, for example, converting street lamps into photovoltaic power sources which supply light, and at the same time supply energy to a city's power grid. Photovoltaic devices differ from solar panels, in that the power they generate can either be stored in batteries or added to the mains power grid [20]. In addition to this variation on solar energy generation, there are items of street furniture available that are wind-powered, or designed as photovoltaic-windpower hybrids; these items, according to their designers, optimize the use of renewable energy sources and also guarantee a power supply through two different sources of energy. Another new power source that is being explored for street furniture is geothermal energy, which arises from the subsoil; heat-controlled benches, for example, can be connected through bore holes in the earth to a geothermal system which produces a comfortable seating temperature [21].

Seating

Seating facilities, such as benches, should be integrated within the outdoor spaces of the city wherever people wait, meet, or socialize; in public squares, for instance, they should be coherent with other elements, so that when seats are not in use they do not create a sense of isolation or emptiness. Being able to sit within a city landscape provides an opportunity to pause, and also a tactile and more intimate contact with a place than one has when standing or walking [22].

Location

The best locations for benches are places where there is heavy pedestrian use: retail shopping corridors, transit stops, plazas, spaces outside cultural institutions, etc; poor locations are areas where there is little activity, such as in front of offices that close early or buildings without windows, and in spaces hidden from view or far away from active areas [23]. In high-activity areas there should be access to seating every 60m (or every 50m if there is intense activity) [24]; there should also be rest facilities at regular intervals of 100-200m, with the needs of the disabled kept in mind [11].

Benches should be set up near other amenities such as bus shelters, kiosks, newsstands, waste receptacles, telephones, etc., and arranged where there is protection from the wind and to allow a choice of sitting in the sun or shade. Moveable chairs or lightweight benches allow the widest range of choices. They should be separated from the walking path by a space of 30-60cm, and with the seats level and having a maximum lengthwise slope of 4%; if the slope is greater, the ground should be built up to create a level surface [25].

Benches set at right angles to each other create situations conducive to socialization [26], whereas if they are lined up in a row they prevent group conversation (they could be lined up in areas with an exceptional view, or where there are many people passing by). In addition, they should not be set up directly facing each other (unless they are being used to play games), because people are reluctant to make direct eye contact with strangers and will turn or sit sideways to avoid it. If pairs of benches are set up at a 90-120□ angle, the space they create facilitates both socialization and sitting alone. There should also be appropriate space left beside and in front of benches to accommodate wheelchairs and walkers for the disabled without hindering pedestrians [27].

Design

In designing benches, considerations of comfort are important, but the level of comfort provided depends on how they will be used: on a shopping street, for example, people may stop briefly to rest, so comfort is less important than it is, say, in a park, where people may spend half a day in seating areas. There may also be other things to consider: for example, large slats in benches might be less comfortable than small slats, but may be more appropriate in areas where teenagers tend to sit on the back rests of the benches. In general, the selection or design of benches should take into account all relevant conditions of use.

Appearance is another important consideration: benches should be designed to fit in with their surroundings, because a bench functions like an extension of the property it adjoins, and gives a street a sense of ownership. As a result, it is more likely that business owners will take responsibility for the care of the bench and the street, and this, on a more general level, will contribute positively to the area's safety and security [27].

Armrests and back support are normal features of seating; these provide assistance to the elderly. Back supports, contoured seats and arm rests provide comfort when sitting and support when getting up and down; these are important design elements, especially in areas where people sit for longer periods of time. Adding armrests in the center of a bench should be considered in cases where it is desirable to prevent people from sleeping on benches. When space is available, steel seating is occasionally used in circular or curved forms to give an organic feel to an area [28].

The design options for a site can be increased if the types and heights of seating (stairs, benches, seat walls, etc.) are varied [18]. The length of seating elements should be 60cm per person. For benches, the recommended height of the seating surface is 42.5cm, with a width of 35-40cm and the back rest extending for 50cm. Benches with arm and back rests provide support to users' bodies; to provide appropriate back support for users, the back rests should be 32.5-37.5cm in height. Between the seating surface and the back rest there should be a 95°-

100° angle, and the seating surface should slope back at an angle of 10°. Walls to be used as seat walls should be 37.5-60cm in height (42.5cm is recommended); and the depth of a seat wall should be at least 37.5cm if access is only available from one side, or 75cm (90cm is recommended) if it is available from both sides [25].

Moveable chairs to be used in open spaces are an alternative to benches and other seating elements; while these may not be ideal in every situation, they can be quite useful, as they are more comfortable than benches. While their cost varies, they are also less expensive: approximately ten moveable chairs can be purchased for the cost of one bench. In addition, users can arrange moveable furniture that they find at a site in any way they wish in order to sit closer together or further apart, in the sun or under shade.

Disabilities

For users with disabilities, some seating should be located near public toilets and telephones. In general, rest areas with benches should allow at least 1m of adjoining space to accommodate wheelchairs [14]; for the same reason, table surface heights should be within the range 0.75-0.90m, and the minimum height under the table should be 0.6m [11].

Waste Receptacles

Location

The most common error in placing waste receptacles is putting them wherever there is an empty space, instead of in places where people will use them. If they are not placed appropriately they will remain empty while the surrounding space becomes dirtied: people will not change their habitual walking path to find a trash can, so the proper placement of these items of furniture is crucial.

Waste receptacles should be highly visible and accessible in order to minimize littering. They should be located where they are most likely to be used, in crowded areas like busy intersections, close to crosswalks, beside take-out food shops or vendors, at bus stops, in plazas, outside building entrances such as offices, department stores and homes, and near other items of street furniture like benches, seats, shelters, walls, fences and telephones [27]. Receptacles should be situated far enough away from seating areas to minimize the unpleasant effects of trash odors and insects on users; and their fronts should be set back at least 30cm from walkways [25].

The number of waste receptacles to be provided depends on how many people use an area, how much litter is produced in the area, and on how efficient maintenance and sanitation programs are. On streets in the city center, 2-4 waste receptacles per block are usually sufficient: one should be at each end of the block next to the crosswalk, and one or two more in the middle of the block if there are benches or food stalls along it [13]. Trash cans are usually spaced at 30m intervals, unless there is an increased need in a particular area; they are usually single units rather than groups of receptacles [24]. On commercial streets there should be at most one trash can every 60.96m; in areas where there is a lot of pedestrian traffic, the distance between receptacles can be decreased, but additional receptacles should be provided only if maintenance is provided by a private sponsor [29].

In some isolated natural areas trash cans may not be necessary, or if they are, they should be selected from a determined range; their location should not detract from the view and they should not have unnecessary decorations that clash with other items of street furniture. In these areas, discreet cigarette stubbers should be provided instead of ashtrays [28].

It should also be taken into consideration that waste receptacles will also be used at night in parks and squares; for this reason, they are most appropriately placed along lit-up pedestrian walkways or other lighted areas.

Design

Waste receptacles come in numerous forms, and may be designed with open, half-open, or closed tops, and mounted on colums (lighting, signs etc.), walls or vertical surfaces, free standing, fixed to the ground surface, moveable (generally for temporary use), built-in bins (benches, seats, walls etc.) [30,31]; however, appropriate receptacles should have certain qualities, of which the most important is that they should clearly look like places for discarding litter; if they blend in with the environment too well they will become unrecognizable. While they should certainly be compatible with other street furniture and with the local architecture, their function and purpose must be clearly recognizable.

Another important quality of a good waste receptacle is ease of use; the following characteristics are the most desirable. First, people should be able to dispose of waste without having to touch the receptacle or open a door to use it. Trash cans are most easily accessible when they are about 100cm in height with the receptacle at most 40cm from the outer edge [19]. The opening of the receptacle should be large enough for litter to be put in, but the size depends on the context; in parks, for example, people may dispose of larger items, so the opening should be bigger. In general, it should be at least 25cm wide, so a folded newspaper or take-out food container can be put into it; if the opening is too small, items can get stuck in it and litter will be strewn in the vicinity of the receptacle. The opening should also be at most 90cm above the ground to facilitate use by the disabled.

The size of the receptacle should be determined by its expected use and how often it will be emptied. Well-organized public spaces have several smaller receptacles that are emptied often, rather than one large receptacle that is only emptied at long intervals. In most areas, containers with a volume of 114-190 litres are sufficient. Wheelie bins are suitable for areas where there is a high volume of trash, if they can be emptied easily by collection equipment. In shopping centers, 120-litre wheelie bins with a framed cover over the top are required; while in town and district parks, 240-litre or 120-litre wheelie bins should be provided, depending on user needs [24].

Waste receptacles should be sturdy; to bear the forces of activities such as sitting, leaning, etc., that may occur beside or on top of them [13]. They should especially be able to contain all kinds of chemical and flammable materials, and should also be robust enough to withstand the impact of children climbing on them or bicycles and other motor vehicles striking them [31]. If they are attached to streetlights they should not obstruct the flow of pedestrians, and should be painted in contrasting colors to assist those with low vision [32].

It is important to know how often waste receptacles will be emptied. In areas where the receptacles are being inspected regularly, they can be lined with plastic bags. If it is expected that the receptacles will be emptied daily, they may be lidless; otherwise there may be a need for them to have a hinged lid resistant to outdoor weather conditions. Lids are also important for receptacles outdoors, to prevent the garbage from getting wet. In addition to having the waste from them collected, the receptacles themselves should also be cleaned from time to time.

The materials used for waste receptacles should be durable and resistant to vandalism, fire, rust, and stains. Among the choices of materials available are: enamels, which are graffiti- and stain-resistant, but may chip easily; wood and rough-textured or porous surfaces, which are rust- and stain-resistant but easily vandalized and hard to clean; aluminum, which generally serves its purpose but may be vandalized to be used as scrap metal in some areas; and plastics, which may be problematic if they are solvent-sensitive, especially in parks, where discarded barbecue lighter fluids can erode them. Galvanized steel is a very durable material that can be used for receptacles in aeas where they may be heavily abused [27].

In view of landfill costs and the currently growing acceptance and understanding of recycling, it seems likely that the need for receptacles for recycling will increase in the future.

Signage

Signage systems play a number of important roles: they provide information and directions for people to find their way around a site, help maintain the site's image and coherence, and encourage learning.

Location

The location of signs significantly reinforces their message. Besides conventional signboards, maps, kiosks, and other elements may also function as signs; and placing signs at other site amenities like benches, cafes, restrooms and intersections can help create mini-destinations in larger open spaces [33].

Signage can be used for various purposes: for example, directional signs can help people keep their bearings and give them a sense of security; and in parks where there are large natural areas or no specific destinations, micro-spaces can be created with a sign, a picnic table, a telephone, or other amenities, facilitating social encounters and enabling communication in the event of an emergency. If the areas where signage is used incorporate other activities, these areas can enable visitors to fulfil various needs in one place, and also create a relaxing social environment in which to spend time.

Design

When people know where they are in a space and know how to get to where they want to go, they feel safer. Instead of using restrictive signs warning that something "is prohibited," a more constructive approach should be used: signage should be positive and informative, and give people the confidence to act on their own. Signage can also be used to educate people on safety issues and to encourage them to report on any dubious activities they may see.

There are four basic kinds of sign: regulatory, warning, informational and educational [34].

Regulatory signs are used for traffic control, and include stop and yield, right-of-way, speed limit and exclusion signs. Directional signs, occasionally posted in natural areas in particular, help people keep their bearings and feel located and secure; they make use of landmarks and other places of interest and are required to indicate changes in direction [19].

Warning signs indicate existing or potentially dangerous conditions; they are usually found near intersections, bridges and crossings, and can also indicate changes in the gradient or surface conditions [34]. Warning signs should make use of all the languages in areas where more than one language is commonly spoken, and should be posted at all entrances to limited, controlled, and exclusion areas. The wording should clearly indicate restricted areas, and the signs (which should not be mounted on fences with intrusion-detection equipment) should be posted at intervals of less than 30.48m [35].

Informational signs inform visitors about a site; in this way they serve as outreach tools. Maps at entrances, within the site and at points along pathways can help increase users' knowledge, curiosity and interest about the area. They help visitors guide themselves, and also highlight places of interest. Maps should not be cluttered with too much or unnecessary information; they should clearly define important points.

Educational signs are a pleasurable way to get people interested in unique features of a site. Educational signage should be created with the types of audiences, their interests, and their reasons for visiting the site in mind.

Poor signage can be confusing rather than illuminating; signs should be posted off-site and at entrances, grouped on one support for clarity and ease of maintenance [24] and should include on-site directional, parking, and cautionary signs for visitors, employees, service vehicles, and pedestrians [35].

Signage should be clearly visible. Directional signs should be visible from a distance of at least 20m: for clear visibility, the sign should not be further away than 47m or closer than 6m. The design and placement of signs for pedestrian use must take into account such matters as their visibility, the size and proportions of the writing, and the relation between the writing and its background. The writing on signs should be complemented by standard graphic symbols to help people; directional arrows should be clearly marked. International symbols should particularly be utilized in outdoor areas used by foreign visitors.

Different materials can be used for signage: carved wooden signs are attractive in natural environments, but they are not resistant to theft and damage; metals can become bent and rusted; plastics can be broken or burned. The most damage-resistant materials are concrete and stone.

All signage should be accessible, and durable in the long term (at least 7-10 years). In areas where there is a risk of vandalism, signs on two supports should be used rather than on single supports. Signs should be easily replaceable in case of damage; and the care of signage should not be neglected.

Shape

The octagon is used only for stop signs [34]; information signboards should be rectangular, warning signboards triangular, and interdictory signboards circular [11].

Colour

There should be a strong color contrast between the text and its background, and between the sign and the surface or background against which it is seen; black text on a white background, for example, provides a good color contrast. Clashing colors, such as green text on a red background, should be avoided. The surface behind the sign should contrast with the sign to make it stand out; or if this is not feasible, the background within the sign should be increased in size [19].

Lettering

The size of the letters used on signs should be proportional to the reading distance: The character width-to-height ratio should be between 3:5 and 1:1, and the character stroke width-to-height ratio should be between 1:5 and 1:10. The letters and graphic symbols should be raised at least 1mm from the background, so the visually impaired can read them by touching them. The smallest letter type should at least 15 mm, and there should be normal spacing between words and letters.

Disabilities

The symbol for amenities for the disabled is composed of a wheelchair figure with a square background or border; the figure is distinguished from the background by contrasting colors, most commonly white for the figure and blue for the background, and the wheelchair figure should always be seen as facing right.

Signs should be accessible to wheelchair users; permanently fixed street furniture will also clearly provide more wayfinding cues for the visually impaired than landmarks [36]. Signs should not be placed behind glass because reflection may hinder visibility. Signs placed on the pedestrian path of travel are obstructive; so they should be detectable.

There is no need for large numbers of directional signs; these should be placed at main entrances and doors and where there are changes in direction or level. Fixed signs indicating street names should be placed at a maximum height of 2.50m. Maps and information panels at building entrances, along roads, and on public buildings should be set at a height of 0.9-1.8m; wall-mounted signs should be installed with the centre line 1.4-1.60m from the finished floor level; and overhanging signs should allow a minimum clearance of 2m. Orientation signs and push buttons in lifts should have a text in Braille or in relief for the visually impaired [11].

Signals at crossings may be supplemented with audible or tactile messages to facilitate crossing for all users, including the visually impaired; however, audible pedestrian signals should be used judiciously, because they can create a noise problem [37].

Lighting

Site lighting design focuses on illuminating the environment to anticipate and respond to the needs of users of the site. The lighting elements involved in a site may have diverse functions such as wayfinding, creating social spaces, interacting with the natural and built environments, and meeting security requirements; the overall design must enable all these functions to coexist experience for users [38].

Because of security concerns, there is often a tendency to over-light parks, plazas, streets, and other public spaces; a good design plan should relate lighting to the functions of a particular space in the evening or at night. Site lighting, beyond its practical functions, should be considered in terms of how the types of lights used, their location and their intensity, affect the way a street is perceived and used [39].

Aesthetically speaking, site lighting can add colour and vibrance to an area at night; however, its primary purpose is to provide safety and security after dark. Lighting is especially important at building entrances, intersections, stairs, sudden changes in grade, dead ends, and remote walkways. Areas with high crime rates should be well lit to provide some security for those using facilities in the evenings or at night; but while it may increase people's feeling of safety, lighting may not have an effect on reducing actual crime rates.

Location

Light poles should be coordinated with other streetscape elements, and utility equipment such as pull boxes and underground trenches, both above and below ground, should be coordinated when placing lighting fixtures. Fixtures should not be placed near tree foliage that may block their light; the anticipated height and diameter of the tree canopy should be considered in relation to the height and spacing of lighting fixtures, the need for a certain level of light, and the need for uniformity. The most suitable distance between a tree and a light fixture depends upon the type of tree and on the type of light fixture; if the light from a fixture is blocked because of the existing locations of the fixture and trees, other light fixtures may be added to achieve the required level of illumination [29].

Design

The lighting of outdoor spaces should be designed carefully, taking into account placement, intensity, timing, duration, and color [18]; lighting can be provided by bollards, street lights, faced lights, shop windows, and other elements of the cityscape. In lighting, as in signage there is a hierarchical order of priority: at the highest priority level, activity areas and primary walkways should be lit to become the focus of pedestrian activity after dark; at the lowest level of priority one may find, for example, the decision not to light some areas at all because using them at night would be unsafe or inappropriate.

Site illumination enhances the safety of traffic and pedestrians crossing; it is used to provide warnings about hazards, and helps increase security and reduce vandalism. Within the plan of the site, it gives emphasis to focal and meeting points, and building entrances. Accent lighting can highlight fine architecture or areas of unusual significance or beauty [17].

Lighting fixtures can also be used to highlight trees and give accents to shrubbery; they can also be mounted in trees to light pathways below. When deciding on such usages, planning should take account of which trees are deciduous and which trees retain their foliage [38].

Steps or stairway lighting should provide sufficient light for people to see the stairs and differentiate between risers and treads: their visibility depends on the materials used for the steps, as well as the physical form of the stairs; dark materials require a higher level of light, and a change in color between risers and treads will increase visibility [40].

Quality and Quantity of Light

Appropriate light sources should be chosen, and fixtures should be designed to direct light precisely, with shielding used to prevent light trespass and glare. The intensity and color of the light should complement the elements to be illuminated. High-pressure sodium lighting, typically used in city street-light fixtures, casts a yellowish-orange glow that distorts colors, diminishes visual clarity and undermines the quality of the night-time urban environment; it should therefore be avoided. Metal halide lighting, in contrast, produces a soft, white glow that renders color accurately, provides better visual clarity, and requires less wattage for the same level of visibility. The quality of light is also influenced by the relationship between the brightness of a light source and one's distance from it: light becomes more diffuse further away from the source, so to produce the desired quality of light, the light source should be located within the range of heights specified for a given brightness.

Height of luminaires

The illumination standard set for pedestrian walkways by the Canadian Standards Association, is 0.4 footcandles, at which level a person's face can be identified from a distance of 12-15 metres [41]. Street and road lights are usually installed on 9-15m high poles, spaced 45-75m apart. Fixed light poles should have durable marking strips in contrasting colors, at least 0.3m long, attached along the centre line at a height of 1.4-1.6m to warn visually impaired pedestrians [32]. Lamps supply an average illumination of one-half a footcandle on local roads and in all parking areas, and one footcandle on major roads and in large parking areas. Pedestrian paths require intensities varying from less than one-half footcandle for walkways to five footcandles for building entrances, steps, and intersections. Mall and walkway light poles are 3-4.5m in height. Although the mounting heights of luminaires have increased in recent decades with lamp technology allowing for higher and brighter road lights, this is not particularly beneficial for pedestrians. If the heights of luminaires are reduced and adjusted to a pedestrian scale, more fixtures will be needed, which means the luminaires and poles and their placement can have a more positive effect on the streetscape [39].

Lamp Types and Wattages

Lamps are usually incandescent or based on mercury vapor. Fixtures where the light source is below eye level may be used to illuminate landscaping and pedestrian walks; if these are the main light sources, there should also be peripheral lighting to illuminate the immediate surroundings and create a feeling of security for passing pedestrians [42]. Excessively bright lights and frontal floodlighting should be avoided; lower-wattage light sources should be used instead. The lighting of hardscape elements from a distance can negatively affect night-time vision, and should also be avoided. Up-lighting should only be used where it will not interfere with the pedestrians' vision [15]: when a luminaire's height is lowered, the lamp's brightness must be adjusted so as not to cause excessive glare for pedestrians, but at the same time, the wattage must also be sufficient to adequately illuminate the road [39].

Reducing light pollution

Light pollution occurs when outdoor lighting is misdirected, misplaced, unshielded, excessive, or unnecessary [18]. Light pollution, or the over-illumination of the night sky by electric lights, can be a significant problem in urban areas; it can negatively impact the normal functioning of humans and many animal species. Light pollution can be decreased if designers minimize light trespass off the site, thus reducing night-time sky glow, increasing night-time visibility, and alleviating the negative effects of light pollution on nocturnal environments. Today newer outdoor lighting products have been designed to minimize light pollution; these types of products, located strategically within the landscape, can provide adequate illumination without emitting excess light [43].

Fountains

Location

In open outdoor spaces, water fountains or drinking fountains provide a focal point. Flowing water has a visual and auditory appeal that creates ambience, and the sound of a water fountain can help screen out traffic sounds in seating areas [44]; drinking fountains should be available for functional reasons as well in regions where there are hot summers [45]. Care should be taken in designing water fountains for high-rise office districts, as the buildings can cause air turbulence that blows around the spray from fountains.

Design

Drinking fountains should be designed on a scale appropriate to their setting. These fountains should be accessible to children and people in wheelchairs, as well as standing adults, including those on crutches or using walkers, and the water control should be simple, with little strength required. A well-designed fountain may have a water spigot on the side for filling containers, or washing hands.

Stone, concrete, brick and metals such as copper, bronze, cast iron and steel are suitable materials for fountains [40]. Taps may also be designed with the disabled in mind, at the usual height of 85-95cm. This allows sufficient space for them to be approached by users of wheelchairs, which are usually 85cm in height; the approach should be on a hard surface. The basin should be cantilevered 20-30cm outwards from the wall or support stand, with the drinking spigot or bubbler mounted on the outer edge of the basin. To operate the fountain, a lever-type handle should be located on the side or rim of the basin; foot pedals should only be used together with a lever for controlling the flow [46].

As wind rises, fountain height should be reduced by 10 percent for each 5mph of wind speed above 10mph. For critical situations a wind sensor should switch off the fountain [47].

Types of fountains

Fountains fall into three categories: rising jets, downward falls, and a combination of the two.

Rising jets are often used in public displays, and because they can rapidly change form, they can be interactive, coordinated with music or easily accessible to people; indeed, the water itself may be the spectacle, if it flows in large quantities or if advanced technologies are used. Rising jets can be seen from a distance, and coordinated lighting can change their appearance into what has been described as «never ending fireworks» [48].

Downward fall or "cascade" fountains are found in nature in the form of rivers, streams, waterfalls and rising springs. Throughout history, human beings have contained, diverted, decorated and reduced or enhanced them, and eventually used them as models, supplemented by artificial pumping systems.

An example of a combination fountain is the "splash fountain," in which people cool themselves; even if they were not designed with this in mind, fountains are often used by children for this purpose. Some fountains may be fenced in or built with raised edges to prevent access; others are designed specifically for easy access, and with nonslip surfaces so they can be used safely. In recent years, purpose-designed splash fountains have been created in open spaces in conjunction with public pools, parks, or public playgrounds. These "splash pads" have no standing water, so there is no risk of drowning in them and no supervision is needed.

Bollards

These are vertical barriers and are one of the most unobstrusive ways of preventing access by vehicles enroaching on to pedestrian areas.

Location

The use of bollards should be limited to areas where sidewalk structures, furniture, private property and vegetation are being damaged by vehicles trying to park; this is particularly problematic on narrow streets. Aesthetically appealing bollards may be used in locations to indicate special spaces like shared public paths or streets limited to pedestrians. Lighted bollards can provide additional illumination for pedestrians in median refuges.

The usual lateral distance of a bollard is about 3m from the center of the street. They should be positioned alongside the the walking path so as not to obstruct pedestrians. Bollards should be installed 45cm from the back edge of the curb [28].

Design

The design of bollards should be contextualized within a 'family' of streetscape elements [29]. Their size range is usually 10-25cm in diameter, but decorative bollards may be larger and may differ in shape. In general, design details appear on the sides and tops of bollards, which should be articulated for this purpose; bollards should also be painted in colors that enhance other elements of the streetscape (except for gray) and to assist the visually impaired. Signs and directions may be affixed to bollards so additional signposts are unnecessary, and the bollards may be painted with white bands where appropriate [28].

The spacing between bollards is usually about 90cm, wide enough to allow luggage and wheelchairs to pass between them. The spacing should also be varied to follow the rhythm of the other elements that make up the streetscape – i.e. lighting fixtures, landscaping, etc. In some contexts, an area may be defined or an entryway blocked by a series of bollards. Bollard forms vary from the simple and modern to the decorative and traditional, depending on the surroundings. The simplest bollards are wooden or painted steel posts embedded in the ground; in their most practical form they demarcate parking areas, alleyways, or entrances to significant public spaces. More complex, detailed forms and refined materials are used when the intention is to blend in with the local architecture and other elements of the city and inform local identity; such forms comprise part of the public street furniture and amenities. The materials and finishes used for bollards are determined by their relation to other elements in the space under consideration [1].

Bollards comprise a wide variety of fixed or flexible elements that are used to demarcate spaces, prevent the entry of vehicles, or otherwise protect a space; they can be designed to give way to or hold up against any force. They can be removable but locked by means of locking pins or by sheer weight [14]. They should be used sparingly, and their style should be determined by their location. Thus distinctive old bollards should be kept and renovated, and new ones created in the same style, if it is still appropriate, using moulds; plastic bollards should only be used for temporary purposes. The design concept for an area should be re-thought in order to minimize the use of bollards: to reduce the need for them, reinforced slabs or stronger paved areas, as well as the use of other street furniture items or planting trees should be considered.

Permanent Bollards

Fixed bollards are also known as "static," "architectural" or "permanent" bollards. These are usually concrete, wood or metal, and cannot be moved as they are surface-mounted or embedded in the ground. They should be set 300-450mm in the ground on a concrete or consolidated hardcore base and surrounded by concrete [49].

Rising (retractable) bollards rise up and go back down into the ground again automatically, by means of an electronic, hydraulic or pneumatic mechanism when they are activated by a hand-held remote, swipe card, or other remote device.

Security bollards are designed to withstand heavy impacts. They may be fixed or retractable, and are used to prevent terrorist or criminal attacks.

Removable bollards have a base permanently fixed to the ground, but can be unlocked with a key and lifted out when necessary.

Telescopic bollards collapse concentrically and retract to street level when they are unlocked with a key.

Lay-flat bollards, also known as "fold-down" or "collapsible" bollards, can be unlocked with a key and laid flat on the ground when required.

Bell bollards are short and bell-shaped; these are used in areas with frequent U-turns or tight turning circles, as the slope of the bell enables the wheel of a vehicle to strike it and roll around it.

Temporary & Flexible Bollards

Qwick Kurb is the brand name of a temporary plastic curb, which is comprised of several plastic bollards or paddles strung together.

Planters can serve as environmentally friendly bollards, physically delimiting streets, widening sidewalks and restricting access for motor vehicles. If they are properly funded, managed and maintained, they can enhance the aesthetics of a public space and facilitate a community's acceptance of a new curb or median [50].

Lighted bollards provide illumination in the form of area or marker lighting. "Area lighting" illuminates the ground plane around the bollard, while "marker lighting" provides a glow indicating a bollard's location. Marker bollards are solar powered, so they do not need to be connected to the electrical grid. In cases where a light is required at a low height for visibility, a simple path light on a post may be more useful and user-friendly than a strong, lighted bollard [1].

Public Art

The term "public art" refers to sculptures and other artworks which are situated in public spaces outdoors and are freely and physically accessible by the public. Public art may belong to the community as a whole, but it can also be displayed in private places and serve a smaller community of interested people [51].

Public art complements and enhances the environment and brings public spaces to life; it can range from disparate objects to an entire streetscape [52]. Large-scale works can bring thematic unity to a district or demarcate a gateway to a neighborhood, while works on a human scale can provide points of visual interest for pedestrians passing by. Well-designed public artworks that recognize the local culture as well as broader influences can define a community, serving as landmarks that uniquely express the spirit of a place or highlight some of its functional aspects, draw people towards it, and so contribute to its growth and dynamism [15]. Public art that fits into its surroundings can thus also create a setting for breaking down barriers, generating diverse forms of social encounters and interactions, and inspiring creative expression [53].

Location

Public works of art can serve as focal points on streets and in public areas, especially at key points or intersections where there is pedestrian movement or where people tend to gather, and they can give these spaces a unique character; however, they may not be appropriate for high-traffic walking areas such as pedestrian thoroughfares, unless they function as street furniture. In addition, they can create a special sense of place in less frequented locations.

Design

Works of public art may be permanent, static or based on objects; they can also be temporary, dynamic, or evanescent; however, they should serve as pedestrian amenities [51], and the aim in the design and planning phases should be to integrate them with other elements of the streetscape such as light poles, benches, trash receptacles and utility boxes. Artworks can be centers of focus in parks or plazas, or visual "surprises" that appears as one follows a pedestrian pathway.

The materials used, whether for a modern or a more traditional sculpture will vary depending on what the sculpture expresses, their appropriateness for its composition, and the cost factors involved. In any case, public artworks should be of a reasonable quality as regards their construction, and particularly their surface finish; they should be structurally sound and designed for minimal maintenance and to resist vandalism [24]. They should also be accessible to the disabled and not hinder pathways; some may need visible warning strips around the base for this purpose.

MAINTENANCE

If the types of street furniture and manufacturers involved at a site are consistent, maintaining or replacing worn or damaged furnishings will be less costly [54]; proper installation of items will ensure their durability, and damaged materials should be replaced with more sustainable products wherever possible [43]. The furniture should be easy to repair, and one way to facilitate this is to design it as replaceable modular parts, so that it does remain unusable for long periods pending repairs [16].

Street furniture is often vandalized by burning, slashing, carving, and spray painting, so this should be taken into account in the planning phase; some materials and styles are more resistant to vandalism. The furnishings must be durable and secure, with finishes and coatings that resist stickers, graffiti and bacteria. Thus while wood may be a cheap and aesthetically appealing choice, it is easy to carve, burn, and spray-paint, so it is not particularly resistant to vandalism compared to other alternatives. Recycled plastic is a popular material for street furniture, but it can be burned, carved and painted; wear due to ultraviolet radiation may be another problem (although some recycled plastic products are marketed as UV-stable), so this material may be more appropriate indoors. Steel is the material most resistant to vandalism, but it is usually the most costly: steel meshes and steel bars cannot be spray-painted, burned or slashed easily, and they also have the advantage of good ventilation so they dry rapidly after rain [15].

CONCLUSIONS

Street furniture should be available in proportion to the intensity of activity in a particular area, and carefully placed to create unobstructed paths for pedestrians without creating hazards. Furnishings include benches, waste receptacle, signs, lighting, fountain and other elements that make people feel comfortable. These elements should be coordinated and integrated so that they are both attractive and functional. However, beyond comfort, the aim of these furnishings is to provide a place with character and identity, and to encourage people to enjoy outdoor spaces. Street furniture also addresses specific needs, such as seating and shelter when one is waiting for transportation. If well planned and designed, it can also enhance the visual aspects, image and identity of a site. The quality, organisation and distribution of street furniture reflects the quality of an urban space and can also set standards and expectations for future development [8], as the aim of these furnishings is to combine and coordinate form, scale, materials and placement to create visual appeal, accessibility and safety through understanding the needs of users [55].

Besides its functional aspects, high quality in the designing of street furniture has recently become a focus of attention in urban landscape design, with emphasis placed on the integration of function and aesthetics with new materials and technologies, creative concepts and artistic application in the user-friendly cityscape. Street furniture enables a city to become closer knit as a community, a space where people can gather, share and experience life together. Visually unattractive or poorly planned street furniture defines a city through chaos, a lack of order and harmony, and the absence of community. In recent years, cities have also become very concerned about waste disposal and the stability of the environment, as can be seen in the growing tendency of using recycled street furniture; well designed street furniture enables cities to continually update outdoor spaces sustainably while being environmentally conscious at the same time.

REFERENCES

1. Main BillHannah Gail Greet. Site Furnishings: A Complete Guide to the Planning, Selection and Use of Landscape Furniture and Amenities,John Wiley & Sons; 201013
2. Transport & Regional Affairs Department of the EnvironmentBy Design, Urban Design in the Planning System: Towards Better Practice, Thomas Telford Ltd; 200026
3. American Association of State Highway and Transportation OfficialsTask Force for Roadside Safety Staff. Roadside Design Guide, 4th edition, AASHTO; 20111013
4. Crankshaw NedCreating Vibrant Public Spaces: Streetscape Design in Commercial and Historic Districts,Island Press, 2 edition; 2008187
5. Evyapan Gönül AslanoğluTokol Ayşegül S. Landscape design lectures, METU Faculty Of Architecture Press; 200081
6. Deakin MarkMitchell Gordon, Nijkamp Peter, Vreeker Ron. Sustainable Urban Development: 2The Environmental Assessment MethodsRoutledge Press; 2007365
7. Sanches Maria Gabriela, Frankel Lois. Co-design in Public Spaces: an Interdisciplinary Approach to Street Furniture Development: Design Research Society Conference 2010, University of Montreal, Montreal, Canada, http://www.drs2010.umontreal.ca/data/PDF/105.pdf (accessed January 2013).
8. Carmona MatthewHeath Tim, Oc Taner, Tiesdell Steve. Public Places Urban Spaces: The Dimensions of Urban Design,Published Elseiver Ltd.; 2010196
9. Gehl JanCities for People, Island Press; 2010239
10. Bain LesleyGray Barbara, Rodgers Dave. Living Streets: Strategies for Crafting Public Space,John Wiley & Sons; 201271

11. United NationsStreet Furniture, Urban Design Considerations, Accessibility for the Disabled- A Design Manual for a Barrier Free Environment, Department of Economic and Social Affairs Division for Social Policy and Development,
12. http:wwwun.org/esa/socdev/enable/designm/AD103htm (accessed January 2013
13. BiddulphMike. Introduction to Residential LayoutArchitectural Press; 2007197
14. Traffic Signs, Marking and Street Furniture,http://www.jtc.sala.ubc.ca/reports/manual_for_streets/10%20traffic%20signs%20marking%20and%20street%20furniture.pdf (accessed January 2013).
15. Derek Lovejoy PartnershipSpon's Landscape Handbook, Taylor & Francis; 1997397
16. Site Furnishing, Urban Green Inc., Vancouver, B.C., http://www.urbangreeninc.com/looking-for/site-furnishings/ (accessed January 2013).
17. Toronto (OntTransportation Services, Toronto (Ont.). Urban Development Services. City Planning Division, Beautiful City Sceretariat. Vibrant Streets: Toronto's Coordinated Street Furniture Program: Design and Policy Guidelines, City Planning, Clean & Beautiful City Secretariat and Transportation Services; 200621
18. Simonds John OrmsbeeLandscape Architecture: A Manual of Land Planning and Design,McGraw-Hill Professional; 1998155
19. Calkins MegThe Sustainable Sites Handbook: A Complete Guide to the Principles, Strategies, and Best Practices for Sustainable Landscapes (Wiley Series in Sustainable Design), John Wiley & Sons; 2012421
20. Planning and designing for pedestrians: guidelines. Department of Transport, 2011.http://www.transport.wa.gov.au/mediaFiles/AT_WALK_P_plan_design_pedestrians_guidelines.pdf (accessed January 2013). 88 89,191,192.
21. Prasad Deo, Snow Mark. Designing With Solar Power. Images Publishing Dist Ac.; 2006.
22. Innovation and special features of street furniture, Sustainable thinking, http://www.repsol.com/es_en/energia-casa/conciencia-

sostenible/reportajes/Innovacion-curiosidades-mobiliario-urbano.aspx (accessed January 2013).

23. Dee CatherineForm and Fabric in Landscape Architecture, Taylor & Francis;2001196
24. Street Design ManualFurniture, New York City Department of Transportation; 2009197
25. Canberra Central Design Manual. DS19- Street and Park Furniture and Barbecues, Design Standards for Urban Infrastructure, Urban Services,http://www.tams.act.gov.au/__data/assets/pdf_file/0007/396880/ds19_bbq.pdf (accessed January 2013). 1913 ,19-5,19-8.
26. Site Design Guidelines a vocabulary for Clemson's sense of place, Clemson University, http://www.clemson.edu/facilities/campus-planning/sdg.html (accessed January 2013). 39
27. Loidl HansBernard Stefan. Opening Spaces: Design as Landscape Architecture,Birkhauser Verlag AG; 2003132
28. Madden KathyStreetscape: A Guide to the Design and Management of Pedestrian Amenities in Downtowns and Neighborhood Commercial Districts, Project for Public Spaces, Inc., 1987
29. Public Space Design GuidelinesFurniture, London Borough of Richmond upon Thames Civic Centre, richmond.gov.uk/psdg_chpt5.pdf (accessed January 201350
30. San Francisco Planning Department, Street Lighting, http://www.sfbetterstreets.org/find-project-types/streetscape-elements/street-lighting/(accessed January 2013).
31. Littlewood Michael Landscape Detailing 3, Routledge; 1997109
32. Harris Charles WDines Nicholas T., Brown Kyle. Time-saver standards for landscape architecture: design and construction data, Second edition, McGraw-Hill publishing company; 1998240248
33. United Nations. Obstructions, Urban Design Considerations, Accessibility for the Disabled- A Design Manual for a Barrier Free Environment, Department of Economic and Social Affairs Division for Social Policy and Development, http://www.un.org/esa/socdev/enable/designm/AD101 .htm (accessed January 2013).

34. Project for Public Spaces. Creating Park Signage; 2009. http://www.pps.org/reference/signage/ (accessed January 2013).
35. Flink, Charles, Olka Kristine, Searns Robert M. Trails for the Twenty-First Century: Planning, Design, and Management Manual for Multi-Use Trails,Island Press; 200188
36. Providing Protection to People and Buildings (Risk Management) FEMA; 2004241
37. Burton ElizabethMitchell Lynne. Inclusive Urban Design: Streets For Life, Architectural Press; 200691
38. Zegeer Charles VPedestrian Facilities Users Guide: Providing Safety and Mobility,U.S. Department of Transportation Federal Highway Administration, Diane Publishing; 200291
39. Ramsey Charles George, Sleeper Harold Reeve. Architectural Graphic Standards: Student Edition, John Wiley & Sons; 2011. 358
40. Project for Public Spaces, Lighting use & design, http://www.pps.org/parks_plazas_squares/info/amenities_bb/streetlights (accessed Jauary 2013).
41. Hopper Leonard JLandscape Architectural Graphic StandardsJohn Wiley & Sons; 2007257
42. Project for Public Spaces, What role can design play in creating safer parks, http://www.pps.org/reference/what-role-can-design-play-in-creating-safer-parks/ (accessed Jauary 2013).
43. H Wood, . Paul, . Paul Wood (AuthorSite Design, Kaplan AEC Education; 2004. p108.
44. Cook Thomas W., Vanderzanden Ann Marie. Sustainable Landscape Management, Design, Construction and Maintenance, John Wiley & Sons, Inc.; 2011. 28 70.
45. Marcus Clare CooperFrancis Carolyn. People Places: Design Guidlines for Urban Open Space, John Wiley & Sons; 199850
46. Marcus Clare CooperSarkissian Wendy. Housing as If People Mattered: Site Design Guidelines for Medium-Density Family Housing, University of California Press; 199229
47. Nelischer MauriceThe handbook of Landcsape Architectural Construction, Volume two, Amer Society Landscape; 198890

48. Littlewood MichaelLandscape Detailing 4, Routledge; 200187
49. HIrst Bryan RFountains, Master Thesis. The faculty of the school of architecture, Kingston University, London, England; 1997
50. Littlewood MichaelLandscape Detailing 1, Routledge; 1993166
51. Transportation Alternatives (Organization)Rethinking bollards: how bollards can save lives, prevent injuries and relieve traffic congestion in New York City,New York, NY: Transportation Alternatives; 20075
52. Coutts GlenJokela Timo. Art, Community and Environment: Educational Perspectives, Intellect Books; 2008147
53. Cartiere CameronWillis Shelly. The Practice of Public ArtRoutledge; 20089
54. Worth Margareth. Creating significance through public places art an inclusive +interdisciplinary practice, Public Art and Urban Design. Interdisciplinary and Social Perspectives Waterfronts of Art III, web version on the w@terfront nr, 4.http://www.ub.edu/escult/epolis/WaterIII.pdf (accessed January 2013). 51
55. LaGro James ASite Analysis: Linking Program and Concept in Land Planning and Design,John Wiley & Sons; 200163
56. Gu Ji YongAdvances in Affective and Pleasurable DesignAdvances in Human Factors and Ergonomics Series), CRC Press; 201222

Chapter 5

Durable and Sustainable Road Constructions for Developing Countries

A.A.A. Molenaar

Road Engineering, Faculty of Civil Engineering and Geosciences, Delft University of Technology, the Netherlands

ABSTRACT

This paper discusses the possibilities to build durable and sustainable pavement structures in developing countries. Attention will be paid to geometric design aspects which have a significant effect on pavement life. Following this attention will be paid to the importance of controlling wheel loads rather than axle loads. Also the importance of controlling tire pressures is discussed. The paper continues with a discussion on the importance of proper construction. Finally it is shown that re-use of construction demolition waste and hot recycling is very well possible and results in good quality materials which can be used to build durable and sustainable pavement structures.

INTRODUCTION

When only limited resources are available to build and maintain road networks, then one should use the available resources in the most optimal way. In terms of pavement design and construction this means that pavements should be designed and constructed in such a way that they can sustain large numbers of heavy traffic loads for a long time without showing a significant amount of damage. This does not necessarily imply that one is talking about thick pavement structures. No, it simply means that one has to do things right and one has to anticipate in a proper way on things that cannot be controlled

Overloading for example is a serious problem in almost all developing countries. If one wants to prevent development of early and excessive pavement damage, then one has two options. The first option is to enforce axle and wheel load regulations or, if that is not possible for one reason or another, design and construct pavements in such a way that they are able to carry these severe overloading's. This latter option however is a costly one while the first option does not necessarily need to cost a lot of money. So why do we not control axle and wheel loads? To a large extent this has to do with the fact that people usually want to exploit whatever that is available to the maximum possible limit without too much concern for the consequences. This later is also true for the environment. We are exploiting Mother Earth to a maximum and we seem not to bother too much about the consequences. We use virgin, "new", materials mostly in an ineffective way and we throw away what we call trash without realizing that this so called "trash" contains very often very valuable materials that can be used again quite easily and quite effectively. Given the fact that well designed and constructed pavements will limit the need to use scarce resources, and given the fact that taking care of the environment and the necessity to recycle as much as possible so called "waste", this paper therefore discusses how long life pavements can be built using recycled materials. Attention will be paid to geometric design and wheel load issues as well as construction quality. Furthermore attention will be given to recycling of asphalt mixtures and construction and demolition waste. It will be shown that materials with excellent characteristics can be produced using recycled materials.

GEOMETRIC DESIGN ISSUES RELATED TO PAVEMENT LIFE

The geometric design of a pavement can have significant effects on its pavement life. It is a well-known fact that the pavement structure at steep gradients and sharp curves suffers from much higher stress condition then the pavement which is laid in a straight section with no longitudinal gradients. Steep gradients and sharp curves result in additional shear stresses which can be the cause of accelerated deterioration especially if the surface layer material is not strong enough or is too aged. Next to that, narrow transverse profiles can also be a reason for rapid deterioration because narrow traffic lanes cause traffic to channel and narrow lanes causing pavement edge loading conditions to occur. Furthermore too narrow lanes have a negative effect on the moisture distribution in an area which is subjected to heavy traffic loads. Both aspects will be discussed in greater detail hereafter.

Narrow Lanes and Accelerated Damage Development

On narrow pavements, the heavy loads come very close to the pavement edge (Figure 1). It has been shown that the tensile strain at the bottom of the asphalt layer of a narrow lane can be a factor 1.05 to 1.2 higher than the tensile strain at the bottom of an asphalt layer of a wide lane because of these edge effects (Molenaar 2007). A 1.05 higher tensile strain means a reduction in pavement life with a factor of about 1.2 while a 1.2 higher strain means a reduction in pavement life of about 2.07! Furthermore too narrow traffic lanes will result in channeling of traffic. If however traffic lanes are wide enough, traffic will show some degree of lateral wander. This lateral wander implies that the maximum tensile strain does not always occur at the same location.

Lateral wander is therefore beneficial to pavement life. Depending on the lane width and the stiffness of the pavement, the lateral wander factor which can be taken into account in fatigue analyses can take a value between 1 and 2.5. All in all this means that a pavement with narrow lanes may show a pavement life which can be 5 times shorter than a pavement which is exactly the same but has sufficient wide lanes!

Figure 1: Edge loadings occurring on narrow traffic lanes.

Narrow Lanes Are Influencing Moisture Conditions under the Pavement

It is a well-known fact that moisture has a significant effect on the bearing capacity of the subgrade soil and the unbound subbase and base. The moisture content under the core of the pavement structure is more or less stable, depending on the height of the groundwater level. The moisture content at the edges of the pavement can however be rather variable because of infiltration and evaporation of rain water (see Figure 2). In case of narrow traffic lanes, not only the outer wheels are close to the pavement edge but they are also standing on an area which might have a reduced bearing capacity in the wet season. This situation is further worsened when the side drains are not functioning properly. This again shows that too narrow pavements might deteriorate faster than pavements with sufficient width.

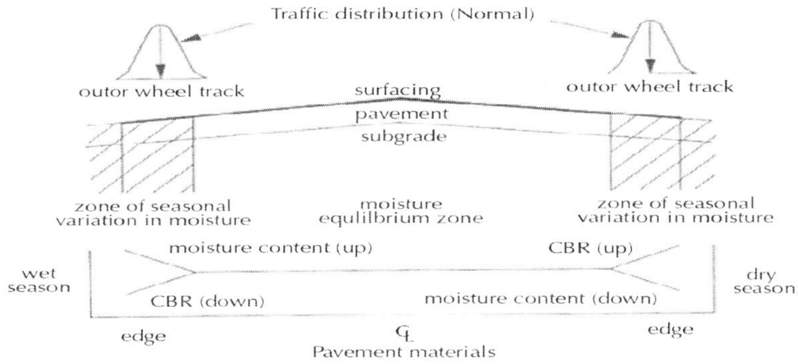

Figure 2: Influence of pavement width on moisture variation in critical areas.

AXLE AND WHEEL LOADS

Overloading is a serious problem in almost each and every developing country. Two problems however are quite often overlooked which are the very high wheel loads and very high tire pressures that regularly occur. One should realize that an axle load which is within the legal limits still might be very damaging simply because of the fact that the axle load is not equally shared over the wheels of the axle. Some examples of very poor distributions will be discussed hereafter. Axle load surveys that were performed in a country in West Africa in the early 1990's showed e.g. that especially the axle loads of timber trucks were very unequal divided over the wheels on either side of the axle (Kumapley and Molenaar 1996). A 124 kN axle load e.g. was carried for 32.5% by the dual wheel bogie at the left side of the axle and for 67.5% by the dual wheel bogie on the right hand side. The load on the right hand side dual wheel bogie was carried for 95% by the inner wheel and for 5% by the outer wheel. This was caused by the fact that a tire with a smaller diameter was placed on the outer wheel. All in all this meant that a wheel load of 80 kN occurred! Given the fact that the legal axle load limit was 100 kN (so 25 kN per wheel assuming a dual wheel on either side of the axle), this 80 kN wheel load was producing approximately 105 times more damage than one would expect! Based on the axle loads however one would expect only 2.4 times more damage due to the 124 kN axle compared to the legal limit axle load of 100 kN. From this simple example it is clear that only axle load enforcement is not enough. Control on wheel loads and equal sharing of the axle load is of equal importance! Another important issue is the tire pressure. The contact pressure underneath a tire is related to the tire pressure and therefore tire pressures should be limited to reasonable levels. The same study in West Africa revealed that the average pressure in the tires of the wheels of the towed axles was approximately 850 kPa while the highest measured value was 980 kPa (Kumapley and Molenaar 1996). The standard deviation was approximately 52 kPa. Pressures measured in the tires of the wheels of the drive axle

However showed an average value of approximately 780 kPa while the maximum values measured amounted 1015 kPa! It will be no surprise that the combination of high wheel loads and high contact pressures is devastating for the surface layer as well as for the pavement structure as a whole. Although it will not be discussed in this paper, it can also easily be shown that trucks with poorly maintained spring-suspension systems are heavily contributing to pavement damage. From a pavement management point of view it is therefore clear that one should pay a lot of attention to the condition of the truck fleet and the overloading conditions. A lot of public money and resources can be saved if the truck fleet is well inspected and axle as well as wheel load regulations are enforced.

INFLUENCE OF CONSTRUCTION QUALITY ON PAVEMENT LIFE

Durable pavements can only be built if the construction quality is up to standard. Skilled and well trained personnel are absolutely needed to "get the job well done". Of course proper functioning equipment is a prerequisite. Figure 3 gives an example of the temperature variation that occurred during the production of an asphalt mixture.

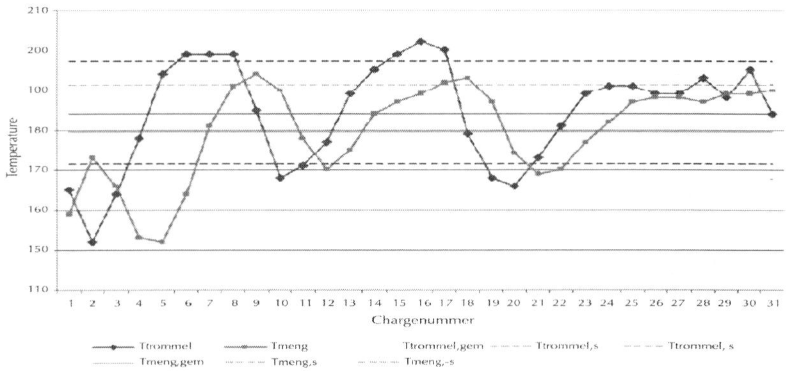

Figure 3: Variation of the temperature in the drying drum (blue line) and mixing unit (red line).

Figure 4 shows the variation in compaction effort that was observed on a regular Paving project in the Netherlands.

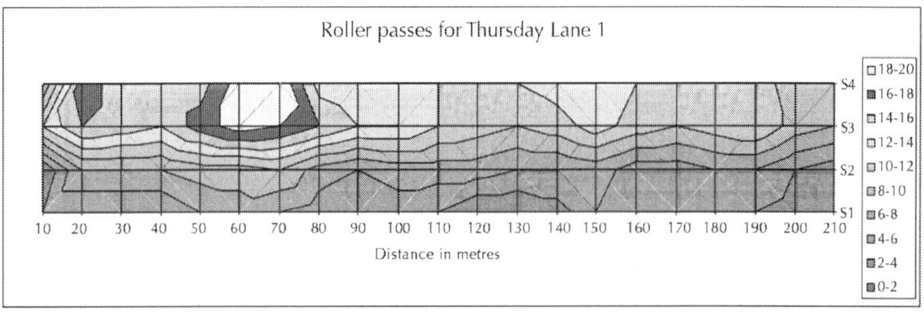

Figure 4: Variation in compaction effort.

From figures 3 and 4 it is quite obvious that significant variations in pavement quality will occur because of variations in mixing temperature and variations in compaction effort. Many more sources of variation can be pointed at. Other examples are segregation of the mixture, temperature segregation, cooling down of the mixture during transportation, excessive moisture in the aggregate storage bins when these are not covered by a shed etc etc. It is clear that one needs not to be a rocket scientist to solve these particular sources of variation. It only asks for good workmanship and the willingness to solve these problems in order to arrive to better performing pavement structures.

RECYCLING

Especially in densely populated areas, recycling of so called waste is a must. Through recycling use of scarce virgin materials is overcome, huge transportation costs related to transportation of virgin materials are overcome, sanitary and environmental problems are overcome etc. Furthermore valuable space which is used for dumping waste can be used for much more useful purposes in case the waste is being recycled. Vast experience with recycling in Western Europe in general and especially in the Netherlands in particular has shown that recycling and re-use does not happen by itself. In order to successfully introduce recycling, the government should take a very active position and it should show a great willingness to co-operate with industry. In the Netherlands, the government has been extremely active in developing and enforcing environmental and recycling related policies. One of the regulations was e.g. making dumping of old asphalt and construction and demolition waste very expensive. Next to that the government was very supportive in stimulating research into e.g. the mechanical characteristics of recycled materials and in stimulating industry to develop and purchase necessary equipment etc. Since the early 1990's recycled asphalt mixtures as well as recycled construction and demolition waste are treated like virgin materials in the standards and specifications.

In this section, 2 important recycling issues will be discussed. The first one is the

Recycling of construction and demolition waste while the second one is the hot recycling of asphalt mixtures.

Recycling Construction and Demolition Waste

In the Netherlands approximately 1 million tons of construction and demolition waste (CDW) is produced per inhabitant per year. In China this number is approximately 0.2 million tonnes per inhabitant. In the Netherlands approximately 95% of the CDW is re-used as base or subcase course in roads. In order to be able to do so, selective demolition is needed. Buildings are first of all stripped from all wood, gypsum plates, asbestos plates etc. In this way, a rather "clean" building is obtained which can easily be broken down. The large blocks are then taken to a crusher to get the desired gradations. During the crushing process electro-magnets are used to take out all metals and metallic parts. Figure 5 is giving an impression of the crushing process.

Figure 5: NW: concrete waste; NE: masonry waste; SW: crushing process SE: crushed material and removal of metal parts.

Mechanical Characteristics of Mixtures of Crushed Concrete and Crushed Masonry

Extensive research on the mechanical characteristics of mixed granulates has been done by van Niekerk and some of his results will be shown here-after (van Niekerk 2002). Figure 6 shows the different gradations that were used in the research program. The codes of the different gradation curves are also used in other figures.

Figure 7 shows the results of monotonic triaxial tests to failure. These tests were performed on large samples having a diameter of 300 mm and a height of 600 mm. Confinement was realized by partial vacuum which implied that the moisture content could not be varied. Figure 7 clearly shows that the cohesion c of the materials increases with increasing amount of fines (UL gradation has more fines than the LL gradation; see also Figure 6). Furthermore the figure indicates the large influence of the degree of compaction on the cohesion. As expected the angle of internal friction is hardly influenced by compaction. Figure 8 shows the dependency of the resilient modulus on the bulk stress as well as the influence of the degree of compaction. Again one will notice the significant influence of compaction on the performance of these materials. The indicator "J" in Figure 8 means that crushing was done by means of a Jaw crusher. In the research different crushers were used in order to determine whether the type of crusher had an influence on the mechanical characteristics. This influence was only marginal and can be neglected.

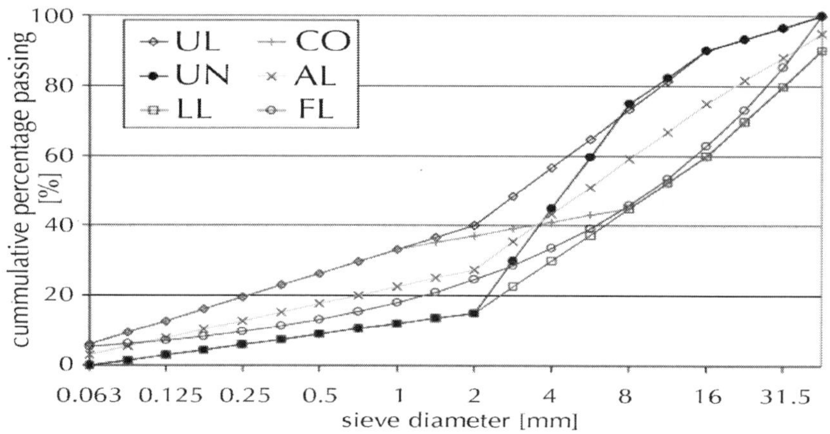

Figure 6: Gradations as used in the research program on mixed granulates.

Finally figure 9 shows results from repeated load triaxial testing which were all performed at a confining pressure level of 12 kPa (this value is representative for the horizontal compressive stress that occurs in granular base courses of pavements as built in the Netherlands). The vertical axis gives the ratio of applied vertical stress to the vertical stress at failure at the same confinement level. As one will observe, a stress ratio of about 0.3 – 0.35 is allowed if one wants to limit the permanent deformation to 1% after 10^6 load repetitions. This level of deformation would be allowable in a perpetual pavement since it would mean 3 mm of permanent deformation after 10^6 load repetitions in a 300 mm thick base course.

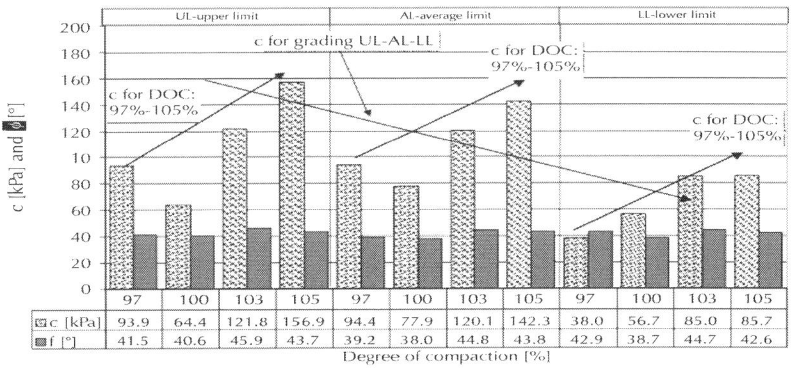

Figure 7: Results of monotonic triaxial tests to failure.

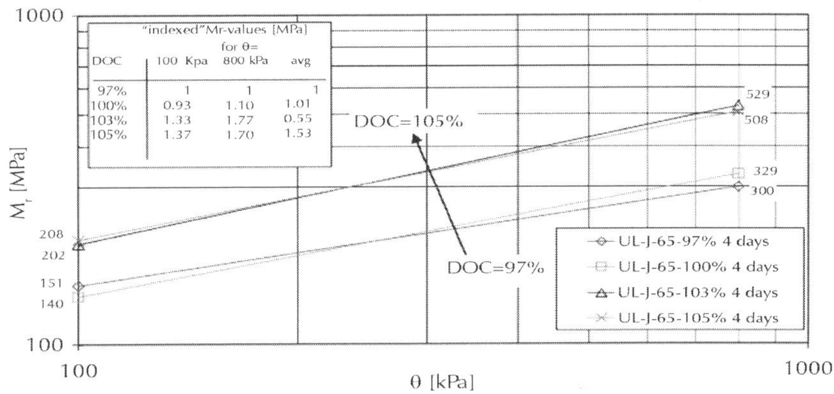

Figure 8: Results of resilient modulus testing.

Research has shown that the resilient characteristics of mixtures of crushed masonry and crushed concrete are as good as those of high quality crushed stone (Araya 2011). However it has also been shown that the σ_1 / σ_{1f} ratios that can be allowed on good quality, well compacted crushed stone are in the order of 0.5. This indicates that the resistance to permanent deformation of high quality crushed stone is somewhat better than that of well compacted mixtures of crushed concrete and crushed masonry. This in turn indicates that mixtures with recycled CDW can be very well used in subbase layers where the stress levels are relatively low. One should be a bit cautious when using them as base course in flexible pavements with a thin wearing course.

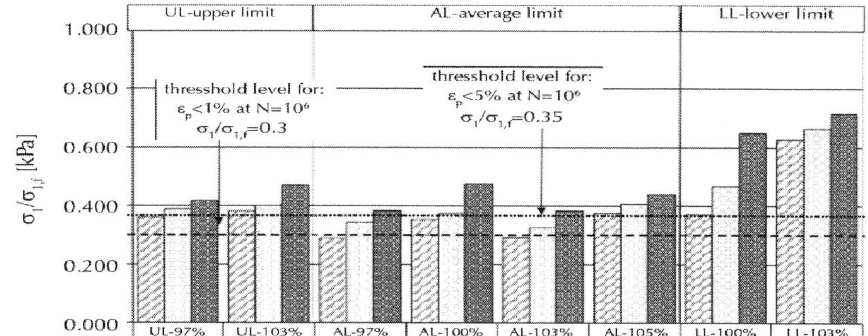

	UL-97%	UL-103%	AL-97%	AL-100%	AL-103%	AL-105%	LL-100%	LL-103%
1%-N=1000000	0.360	0.381	0.289	0.352	0.291	0.376	0.371	0.628
5%-N=1000000	0.388	0.401	0.343	0.374	0.325	0.407	0.467	0.665
10%-N=50000	0.416	0.471	0.382	0.475	0.383	0.440	0.649	0.715

Figure 9: Resistance to permanent deformation of mixtures of crushed concrete and crushed masonry.

Hot Recycling of Asphalt Mixtures

Recycling of reclaimed asphalt (RA) in hot mix asphalt is a well know technology for many years. In the Netherlands approximately 0.2 ton RA is produced per year per inhabitant. In the Netherlands 80% of the RA is re-used in hot mix asphalt. A maximum of 50% RA is allowed in the bituminous base, binder and dense asphalt concrete wearing courses. No RA is allowed in SMA while 20% of RA is allowed in porous asphalt concrete. The most common way to produce recycled asphalt mixtures is to preheat the RA in a parallel drum to a temperature of about 130 ^{0} C. The virgin aggregates have to be heated to temperatures well above 200 ^{0} C to arrive to a mix temperature of about 170 ^{0} C. It is obvious that the temperature to which the virgin aggregates have to be heated is dependent on the amount of RA added to the mixture. The penetration grade of the bitumen to be added is determined by means of the log (pen) rule which is:

$$a * \log(\text{pen1}) + b * \log(\text{pen2}) = (a+b) * \log(\text{penmix})$$

Where:

a = volume percentage of binder coming from the RA

b = volume percentage of new binder

pen1 = penetration of the binder coming from the RA

pen2 = penetration of the new binder to be added

Because maintenance of the existing road network is becoming a much more important issue in the Netherlands than the buil-ding of new roads, there is a tendency to increase the amount of RA in asphalt mixtures. It has been shown by Rasenberg contractors that mixtures made with 100% RA can be made provided that the RA binder is rejuvenated with a special vegetable oil. Furthermore it has been shown that fractionizing the crushed RA and storing it in different bins is important to keep a good control over the quality of the recycled mixture. Figure 10 shows how the various RA fractions are stored separately at Rasenberg contractors.

Figure 10: Storing RA aggregates in different fractions.

Table 1: Binder content related to RA aggregate size

Fraction size [mm]	0 - 2	2 - 5	5 - 8	8 - 11	11 - 16	16 - 22
Mass percentage of total aggregate fraction	22	21	15	18	16	8
Percentage of binder in that fraction	33	25	11	13	13	5

The importance of fractionizing is shown in Table 1 which indicates that the fine fractions of the RA are carrying most of the RA binder. The fractions smaller than 5 mm amount 43% by mass of the total aggregate fraction but they carry 58% by mass of the total amount of RA binder.

Mechanical Characteristics of Hot Recycled Asphalt Mixtures

Table 2 shows some mechanical characteristics as determined for several recycled mixtures made with different amounts of RA. The table also shows the mechanical characteristics of recycled mixtures which also include shredded bituminous roofing material. The table clearly shows that recycled mixtures containing a high amount of RA and shredded bituminous roofing material can be made which comply with the specifications according to the European norms. This in turn implies that good quality roads can be built using asphalt mixtures having high RA contents.

Table 2: Some mechanical properties of hot recycled asphalt mixtures

	E* max [MPa] At 20 0C and 8 Hz	**$\in$ fatigue at N = 10^6 at 20°C and 30 Hz [μ m/m]**
Base course mixture requirements EN	11000 - 14000	80 - 100
Mixture with 100% RAP + resin	15647	101
Mixture with 70% RAP + resin + roofing	10219	139
Mixture with 45% RAP + resin + roofing	10164	139

CONCLUSIONS

Several aspects have been discussed in this paper. First of all it has been shown that a proper geometric design can result in significant longer pavement lives. Especially the pavement width is a factor to consider. Furthermore it has been shown that excessive wheel loads should be overcome. Measuring axle loads only is not enough. Time and money spent on vehicle controls should really incorporate control of wheel loads and tire pressures. It also has been shown that proper control during pavement construction could overcome building in weak spots during construction. Proper equipment as well as well trained and dedicated personnel are key factors in achieving this. Finally it has been shown that re-use of construction and demolition waste as well as hot recycling of asphalt mixtures is an excellent way to reduce the need to use virgin materials and is an excellent way to re-use so called construction waste. All the aspects discussed do not require large investments or fancy equipment. Therefore it is believed that durable and sustainable pavement structures can also be very well built in developing countries. It is however important that the government develops active policies to promote and stimulate recycling of building materials.

REFERENCES

1. A.A.A. Molenaar (2007). "Lecture notes CT4860 Structural Pavement Design; Design of Flexible Pavements". Delft, the Netherlands; Delft University of Technology.
2. A.A. Araya (2011). "Characterization of Unbound Granular Materials for Pavements". PhD Dissertation. Delft, the Netherlands;Delft University of Technology.
3. A.A. van Niekerk (2002). "Mechanical Behavior and Performance of Granular Bases and Subbases", PhD Dissertation. Delft, the Netherlands; Delft University of Technology.
4. N.K. Kumapley and A.A.A. Molenaar (1996). "Axle Load Survey Report". Report to the Commission of the European Communities. Accra, Ghana.

Chapter 6

Case Study of Remaining Service Life Assessment of a Cooling Water Intake Concrete Structure in Indonesia

M. Sight Darmawan, Redo Bayuaji, N. A. Husain, and R. B. Angara

Civil Engineering Diploma Program, Institute Technology Speculum November (ITS), Surabaya 60111, Indonesia

Received 8 August 2014; Revised 10 October 2014; Accepted 10 October 2014; Published 11 November 2014

Academic Editor: Andreas Kaposi

ABSTRACT

This paper deals with the assessment of remaining service life of a cooling water intake concrete structure (CWICS) subjected to corrosion due to chloride attacks. Field and laboratory tests were performed to determine the current existing condition of the structure. Both destructive and nondestructive tests were employed to obtain the parameter needed for the assessment. Based on the current condition and test results, structural analysis was carried out and the remaining safety factor of CWICS was determined. From the analysis, it was found that most concrete elements of CWICS had safety factor greater than unity and might fulfil its intended service life up to the year 2033. However, fewer elements require immediate strengthening to extend their service life.

INTRODUCTION

Corrosion of reinforcing steel due to chloride attack is considered to be the primary cause of concrete deterioration of reinforced concrete structure [1]. This factor combined with poor practice in detail design, bad supervision, and bad construction execution lead to early deterioration of concrete structures. Concrete structures built 30–40 years ago often do not comply with the present day and more modern code requirement for durability. For example, most of the present day concrete codes specify that the minimum cover for concrete structures built in marine environment is 65 mm [2], whereas the corresponding minimum concrete cover during that time is around 50 mm. Furthermore, theoretical foundation of chloride penetration in concrete structure was not yet fully developed and well understood at that time. This lack of knowledge and understanding on concrete deterioration mechanism lead to no intended faulty concrete practices. Therefore, it is not surprising that older concrete structures often has durability problem before their design life has expired.

It is also expected for concrete structures built in a tropical country such as Indonesia to have higher corrosion rate than that of concrete structures built in temperate or cold region [3]. This higher corrosion rates are caused by higher average temperature and higher humidity experience by concrete structures along the years. Furthermore, workmanship and construction practice in Indonesia is not as good as those in a developed country. All of these factors may lead to early deterioration of concrete structures and shorten the service life of concrete structure.

CASE -STUDY

This paper presents a study of remaining life assessment [4] of cooling water intake concrete structure (CWICS) at Indonesia. The study is comprised of field and laboratory test and followed by analytical study. CWICS has been in service for 19 to 33 years and subject to continues chloride attack from nearby sea. Therefore, it almost reaches its design service life of 30 years. In addition, part of CWICS also subjects to high temperature from the discharge cooling water from the factory. This higher temperature can increase corrosion rate of steel rebar in concrete [5]. All of these conditions may shorten the service life of CWICS and endanger the factory operation. CWICS has an important role in gas production factory as it supplies cooling sea water needed by the factory.

At present, some parts of CWICS have shown some signs of damages, such as staining, rusting, cracking, spelling, and delamination of concrete, see Figure 1. These damages indicate that the chlorides may have already penetrated concrete cover, reached rebar level, and accumulated to threshold level chloride concentration to initiate corrosion. The corrosion may have reduced rebar cross-section and lead to a reduced strength capacity of some structural elements of CWICS. If this condition is not rectified soon, it may endanger the whole structure of CWICS and shut down the factory operation. The shutdown of the factory can lead to significant loss of revenue to factory owner.

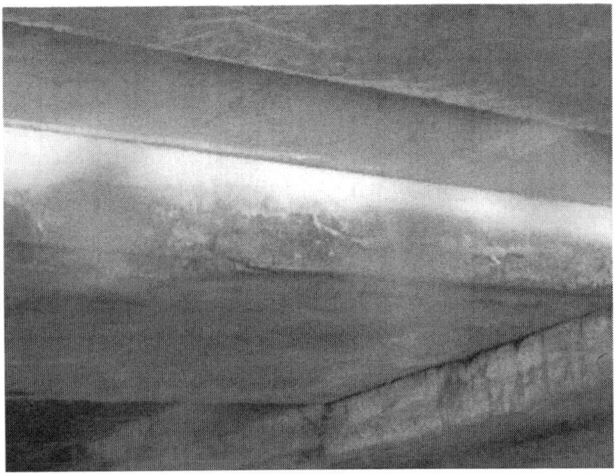

Figure 1: Cracking of concrete plate (slab) of CWICS.

The purpose of this study can be summarized as follows:

- determine current existing condition of CWICS;
- Determine remaining life of CWICS ;(
    - (a) Determine the safety factor of CWICS at year 2013;
    - (b) Determine the safety factor of CWICS at year 2033.

STRUCTURAL CONFIGURATION OF COOLING WATER INTAKE CONCRETE STRUCTURE

Cooling water intake concrete structure (CWICS) is made of concrete structure supported by steel piles. The concrete structure of CWICS comprises plate (slab), beam, and wall elements. A steel frame is installed on the top of CWICS for a crane operation (see Figure 2). In addition, a number of machines for pumping sea water are stationed on top of the concrete structure. Most of these machines run for 24 hours without stopping. CWICS consists of 4 trains, which has almost similar structural configuration. These are trains A/B, C/D, E/F, and G/H, built in 1977, 1982, 1987, and 1995, respectively. These trains were built by different contractors.

Figure 2: Front view of CWICS.

METHODOLOGY

To determine remaining service life of CWICS, the current condition of CWICS needs to be investigated and rate of deterioration needs to be determined. The ultimate goal of this study was to determine whether CWICS can fulfill its intended service life up to 2033 without strengthening. To achieve this goal, the following steps and tests were employed in this study.

- Collect information regarding design criteria from available document and as-built drawing and any changes might occur during service period.
- Determine current concrete density and concrete compressive strength of CWICS.
  - (a) Compression test of core-drilled concrete sample
  - (b) Ultrasonic pulse velocity (UPV) test.
  - (c) Hammer test
  - (d) Porosity test.
- Determine carbonation depth.
  - (a) Phenolphthalein test.

- Determine yield strength of rebar and remaining steel rebar thickness.

  (a) Tension test of rebar samples taken from core-drilled concrete samples.

  (b) Thickness loss measurement of corroded rebar.
- Determine chloride content and pH of the concrete at different depth.

  (a) Chloride content test from core-drilled concrete samples.

  (b) PH test.
- Determine chloride and sulphate contents of seawater.
- Determine probability of corrosion of rebar

  (a) Half-cell potential measurement.
- Structural and load modeling of CWICS using available finite element. Program to determine the internal forces.
- Determine rate of concrete deterioration

  (a) Concrete cover depth measurement.
- Determine present capacity of structural element of CWICS.
- Determine remaining service life of CWICS.

At present paper, only concrete structure of CWICS is considered. Steel piles that supported CWICS will be discussed in another study. It must be mentioned here that, during field tests, factory operation must not be interrupted. Further, safety measure in the studied area was very tight and only limited access was given to do the field test. Therefore, the number and the location of tests performed were rather limited. To compensate this deficient, data interpretation of the tests was combined with engineering judgment to predict remaining service life of CWICS. Due to limited number of data obtained from this study, only deterministic approach was discussed in this paper.

Determine Current Concrete Condition of CWICS

Information regarding the design compressive concrete strength of CWICS can be found in the available as-built drawing and document specification. The specified concrete strength was 28 MPa

With a maximum water-cement ratio of 0.4 and used type II cement. This concrete strength is slightly lower than the present day minimum concrete strength requirement for marine environment of 35 MPa. However, the actual compressive strength achieved during construction was not well documented. Therefore, this data must be obtained by performing field and laboratory test. Four different tests were used to estimate current concrete condition of CWICS. These included compression test of core-drilled concrete sample, hammer test, UPV test, and porosity test. The most accurate method to determine concrete strength is compression test of core-drilled concrete sample. However, this destructive method is very expensive to perform and create permanent defect at the existing structure (see Figure 3). Therefore, this method was combined with nondestructive test, such as hammer and UPV test to get more data for concrete strength indication and homogeneity. Hammer and UPV tests were performed for each location of core-drilled concrete sample and other locations. If the number of data is sufficient, a correlation chart between these tests and compressive strength can be derived. Using this chart, the concrete strength can then be inferred both from hammer and UPV tests. However, as shown later in the next section, a good correlation factor was not always obtained between these tests due to a number of reasons.

Figure 3: Core-drilled sample.

Table 1 gives the number of concrete core-drilled samples for each train. This table shows that more samples are taken from older train than newer one. This approach was employed as older train has shown more sign distress than newer train. The location of core-drilled sample at train A/B is shown in Figure 4. A similar pattern of sampling was also used for other trains. To avoid rebar in the concrete, the location of core-drilled was first checked using rebar detector before any drilling operation commenced. However, out of fifteen core-drilled samples, only ten samples were successfully compression tested and five samples were broken during the drilling process. The broken samples were examined and it was found that cracks were formed in these samples. The core-drilled samples were obtained from the top of CWICS as the access from the other side was very limited and the factory must operate at all times without stopping. The compressive strength of core-drilled samples is given in Table 2.

Table 1: Number of concrete core-drilled sample

Train	Numbers
A/B	5
C/D	4
E/F	3
G/H	3
Total	15

Table 2: Compressive strength of core-drilled sample.

No.	Code	Location	Compressive strength (kg/cm^2)
1	Core 1	Train A/B	222.13
2	Core 2	Train A/B	246.63
3	Core 3	Train A/B	279.66
4	Core 6	Train C/D	408.35
5	Core 8	Train C/D	401.70
6	Core 9	Train C/D	377.76
7	Core 10	Train E/F	373.76
8	Core 11	Train E/F	351.93
9	Core 12	Train E/F	397.65
10	Core 14	Train G/H	411.01

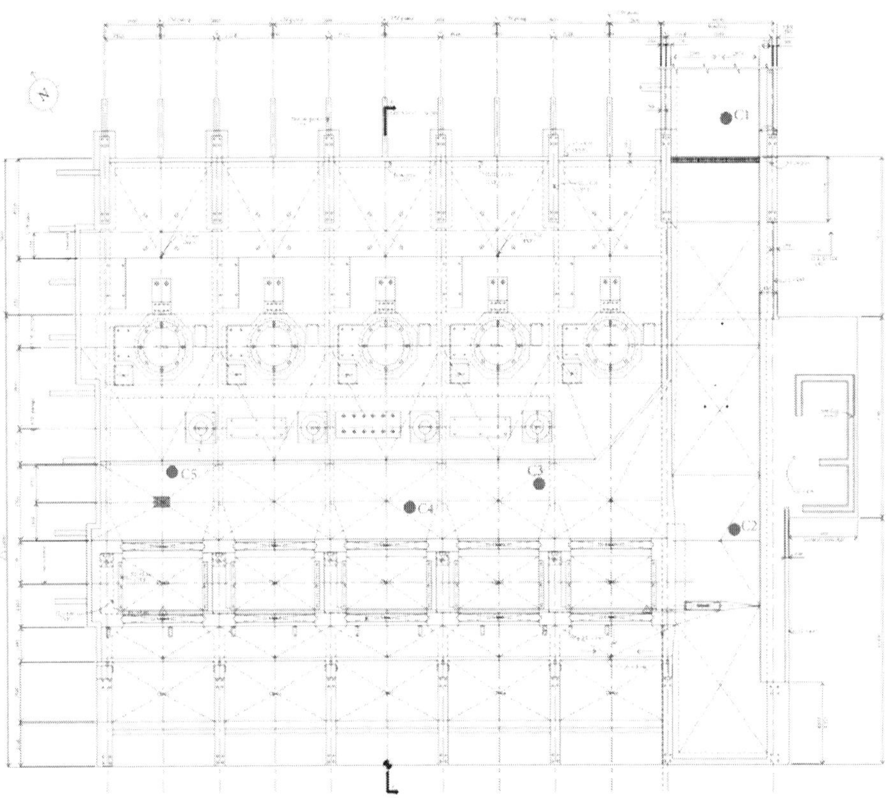

Figure 4: Core-drilled location at train A/B (Core 1/C1 up to Core 5/C5).

Table 2 shows that the compressive strengths of core-drilled samples of train A/B are lower than the compressive strength of core-drilled samples from the other trains. This result may indicate that the concrete at this oldest train has already experienced more strength degradation than concrete at the other trains. The concrete strength at train A/B is lower than the present day minimum concrete strength requirement for marine environment (i.e., 350 kg/cm^2) such as stipulates in [2] and also lower than the specified concrete strength of 280 kg/cm^2 as found in as-built drawing. By comparison, the highest compressive strength was obtained at the newest train G/H at 411.01 kg/cm^2. However, only one core-drilled sample has been successfully tested for this train. In addition to concrete core drill, UPV and hammer tests were performed as shown in Figures 5 and 6. The location of these tests can be seen in Figures 7 and 8, respectively.

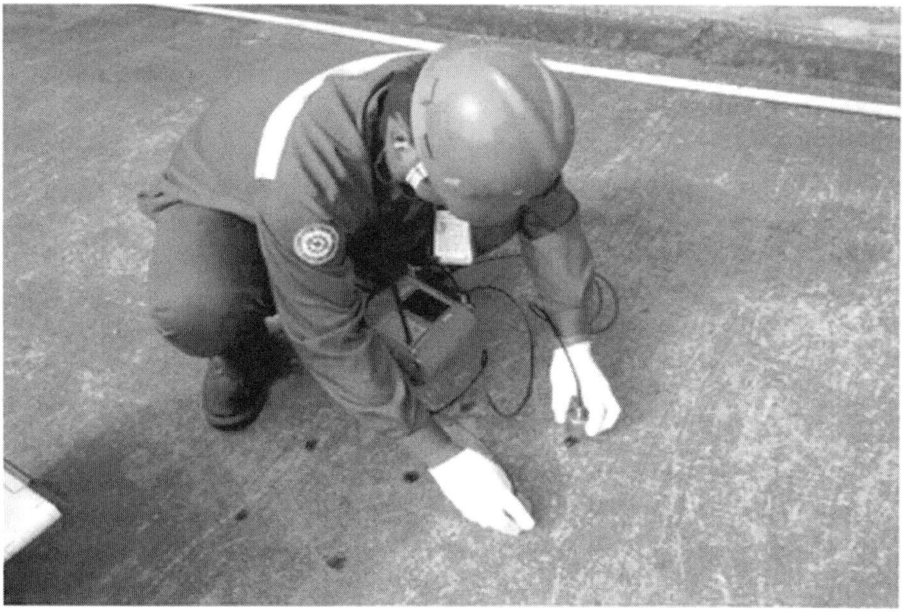

Figure 5: UPV test at train A/B.

Figure 6: Hammer test at train A/B.

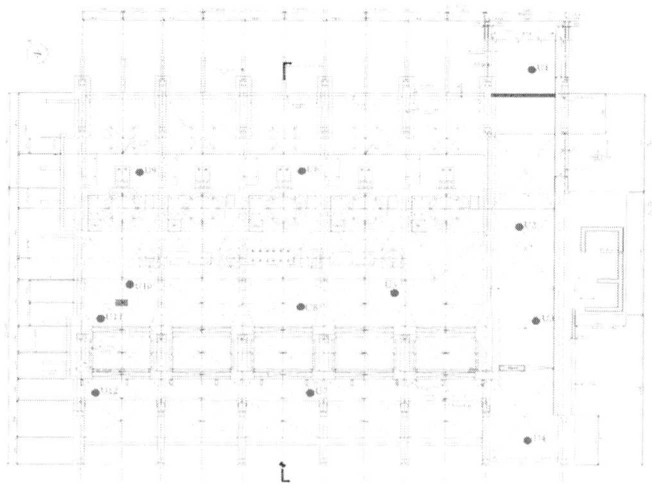

Figure 7: UPV test location at train A/B.

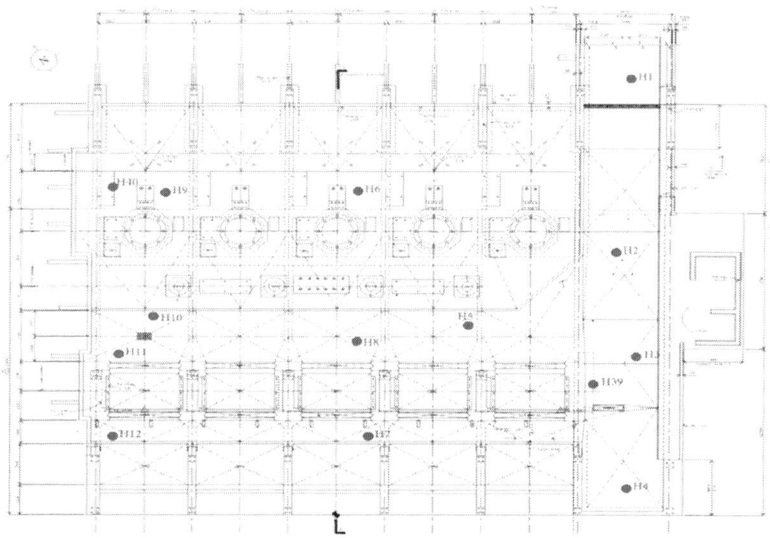

Figure 8: Hammer test location at train A/B.

Table 3 shows the ultrasonic velocity and its corresponding compressive strength for all trains. This table shows that almost all ultrasonic velocities in the concrete fall below 3000 m/s, except the ultrasonic velocity of cores 3 and 8. Based on [6], these low ultrasonic velocities can be classified as doubtful. These low readings of ultrasonic velocity are possibly due to discontinuity that presents in the concrete plate (slab) of CWICS. After a close examination of core-drilled samples in the laboratory, it was found that a 20 mm no shrinking grouting material was laid on top of concrete plate to give additional protection against chloride environment. Because this material and the old concrete below have different properties, discontinuity presents between them. This discontinuity reduces the ultrasonic velocity in the concrete. The ultrasonic pulse may be diffracted around the discontinuities, therefore increasing the travel path and travel time [7].

Table 3: Compressive strength and ultrasonic velocity of core-drilled samples

Number	Code	Location	Compressive strength (kg/cm^2)	Ultrasonic velocity (m/s)
1	Core 1	Train A/B	222.13	1830
2	Core 2	Train A/B	246.63	2338
3	Core 3	Train A/B	279.66	3232
4	Core 6	Train C/D	408.35	2420
5	Core 8	Train C/D	401.70	3002
6	Core 9	Train C/D	377.76	2413
7	Core 10	Train E/F	373.76	1842
8	Core 11	Train E/F	351.93	2263
9	Core 12	Train E/F	397.65	2008
10	Core 14	Train G/H	411.01	2610

Table 3 indicates that core 1 drilled at train A/B gives the lowest ultrasonic velocity of 1830 m/s. This lowest value corresponds with its lowest compressive strength of 222.13 kg/cm^2. Similar trend is also found for train C/D where low compressive strength corresponds with low ultrasonic velocity. However, this trend does not apply for train E/F where low compressive strength gives high ultrasonic velocity. Table 3 also shows that the highest ultrasonic velocity of 3232 m/s is found at core 8 with its corresponding compressive strength of 279.66 kg/cm^2. As each train was built in different years and used different concrete mixes, correlation chart between UPV and compression strength for each train was derived separately. The correlation chart is shown in Figures9, 10, and 11 for train A/B, C/D, and E/F, respectively.

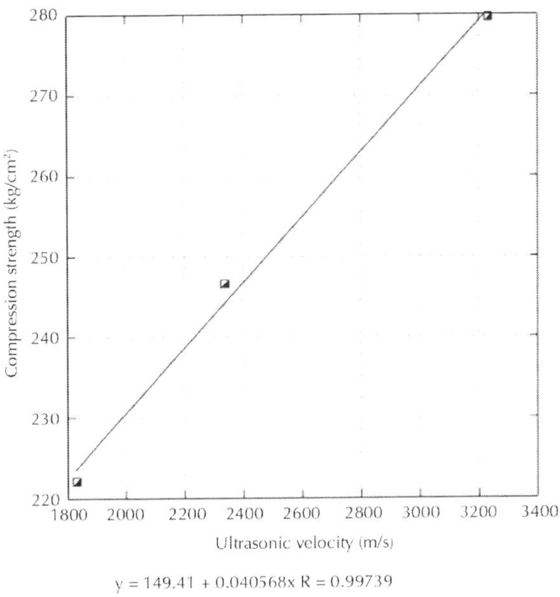

Figure 9: Correlation between UPV and compression strength for train A/B.

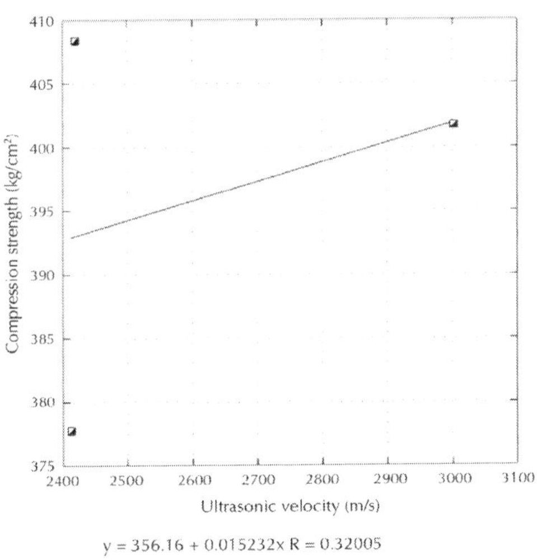

Figure 10: Correlation between UPV and compression strength for train C/D.

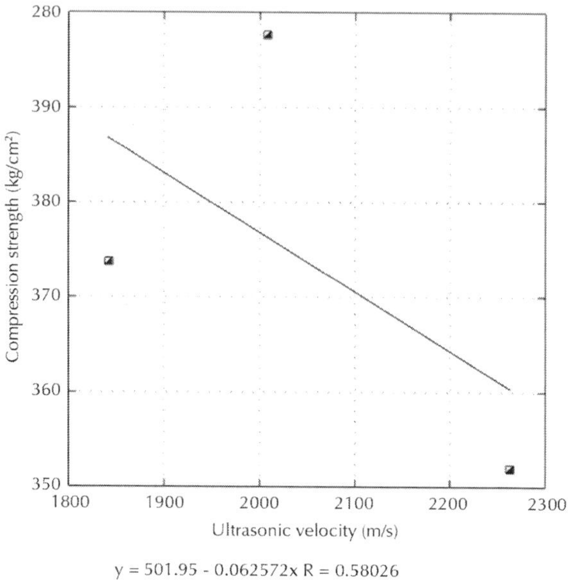

Figure 11: Correlation between UPV and compression strength for train E/F.

Figures 9 to 11 show that the best correlation between ultrasonic velocity and compressive strength is found for samples taken at train A/B, with a correlation factor () of 0.997. On the contrary, Figure 11 shows an opposite trend between these two tests at train E/F, where the highest ultrasonic velocity gives a lower strength. Again this result confirms that nondestructive test results should not be used solely without destructive test as it may lead to wrong interpretation.

Figure 12 shows a correlation chart between hammer and compressive strength at train C/D. It gives a reasonable correlation factor () of 0.72089. However, if all hammer tests for all trains are combined in one chart, the correlation factor between hammer and compressive strength drops to 0.19884 as shown in Figure 13. It must be mentioned herein that, before hammer tests were performed, the hammer equipment was calibrated first using standard anvil from the manufacturer. Further, the concrete surface was first grinded to obtain flat surface. However, the rebound numbers obtained during the test were lower than those available in the literature and also gave lower correlation factor between hammer and compressive strength

[8]. One possible explanation of this condition to occur was that the hammer tests were performed on the top side of concrete plate. As discussed earlier in the section, it was found that, during the service life of CWICS, a 20 mm non-shrinking grouting material was laid on top of the concrete plate. This material does not contain any coarse aggregate and therefore leads to lower rebound number of hammer tests. Hammer test performed on the other elements such as beam and wall elements gave a higher rebound number than that obtained from concrete plate element. However, no concrete drill samples were taken from beam and wall elements as field condition did not allow the drilling process to be executed on these elements.

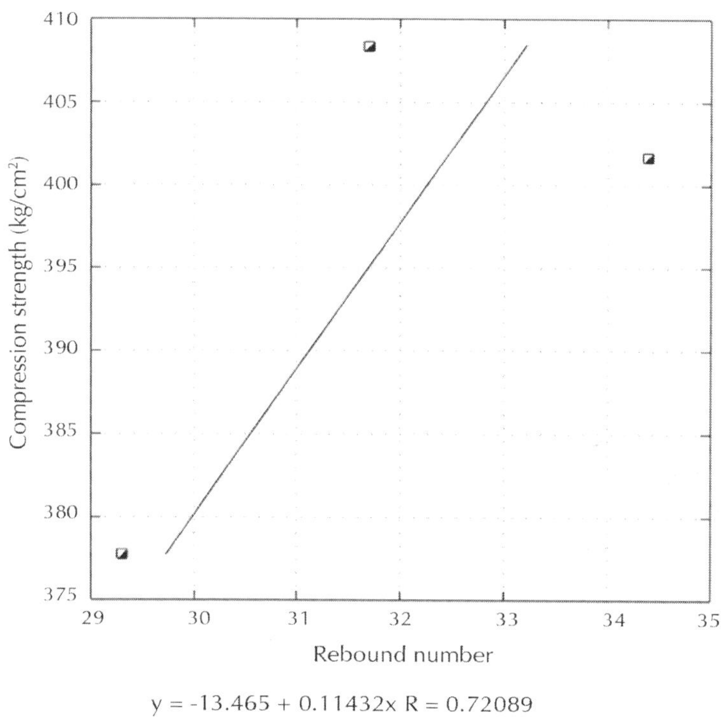

Figure 12: Correlation between hammer and compression strength for train C/D.

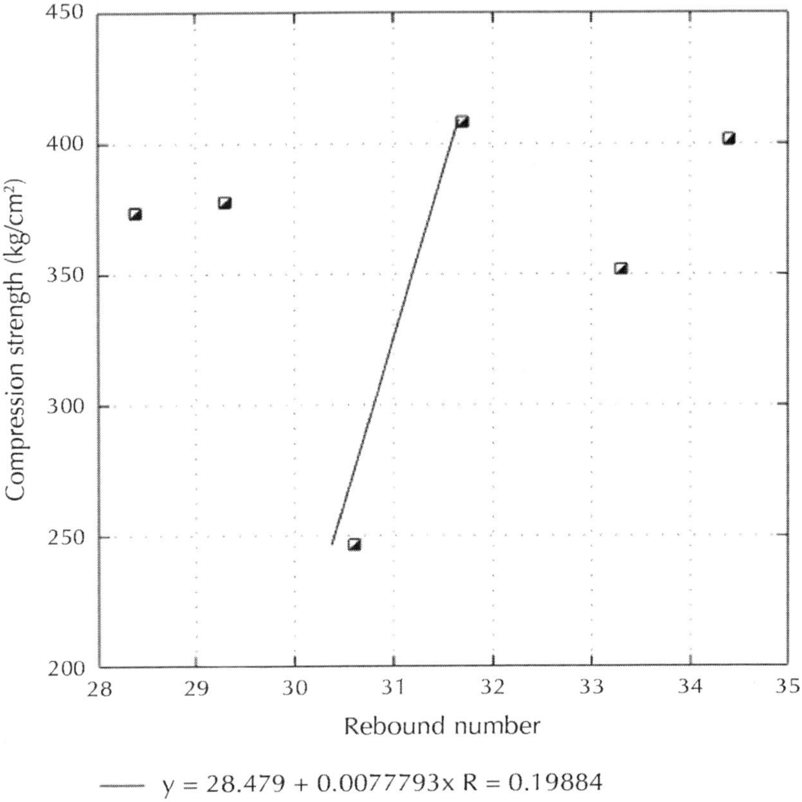

Figure 13: Correlation between hammer and compression strength for all trains.

Concrete porosity is the major factor that influences both strength and durability of concrete structure. Concrete with high porosity has a low concrete strength and low durability. A number of methods can be used to determine the porosity of concrete such as saturation method, helium pycnometry and mercury intrusion porosimetry. For this study, porosity test was performed using vacuum saturation apparatus [9]. The result of this test is presented in Table 4. This table shows that most of the sample has a porosity less than 10%, except for the sample taken from core 1. Core 1 has the highest porosity of 11.5%. This value also corresponds with its lowest compressive strength of all samples. By comparison, core 6 has the lowest porosity at 4.3%, but it gives only the second highest value of all compressive strength.

Table 4: Porosity test of concrete core-drilled samples.

Code	Train	Porosity	Compressive strength (kg/cm^2)
Core 1	A/B	11.5%	222.13
Core 2	A/B	7.8%	246.63
Core 6	C/D	4.3%	408.35
Core 9	C/D	9.8%	377.76
Core 10	E/F	8.3%	373.76
Core 14	G/H	8.4%	411.01

Compared with the available data in the literature [10, 11], the porosity of concrete given in Table 4 is lower for the same concrete strength. The data in the literature shows that for concrete strength of 30 to 40 MPa, the porosity of concrete is found around 15 to 20%. On the contrary, data in Table 4 shows concrete porosity of 4.3–11.5% but with corresponding max concrete strength of only 40 MPa. It must be mentioned herein that all the available data in the literatures was mostly taken at 28–90 days old, while porosity data presented herein was taken after 19–33 years old. It appears that older concrete gives lower porosity than younger concrete but without significant strength gains.

Figure 14 shows the correlation chart between porosity and compressive strength for all trains. Compared with Figure 13, porosity has a better correlation to compressive strength than hammer test, having a correlation factor of 0.54. This result again confirms that destructive test such as porosity test has a better accuracy than nondestructive test such as UPV test. However, porosity test requires the samples to be taken from an existing structure and therefore it is expensive to perform.

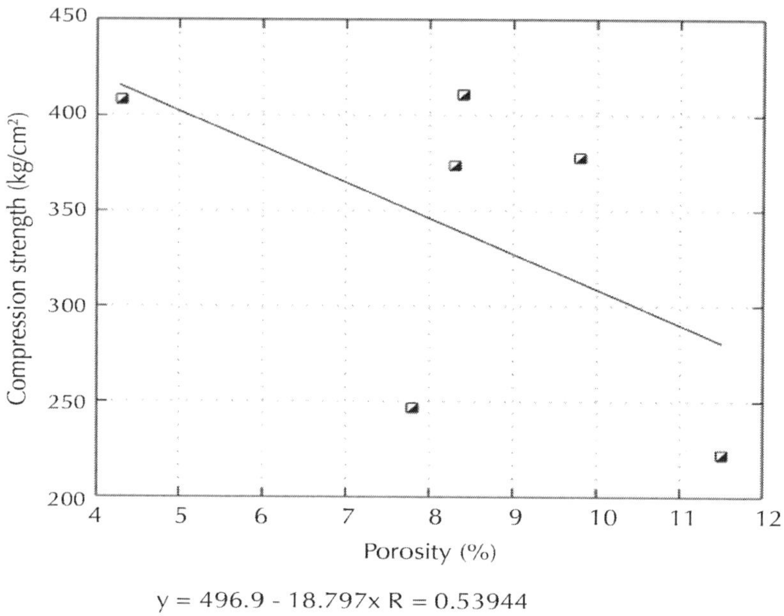

Figure 14: Correlation between porosity and compression strength for all trains.

Determine Carbonation Depth

After concrete core-drilled sample was obtained, the cylinder specimen was then straight away tested for depth of carbonation. Depth of carbonation was checked using a solution of phenolphthalein indicator that appears pink in contact with alkaline concrete with pH values in excess of 9 and colorless at lower levels of pH [12]. This test is most commonly performed by spraying the indicator on freshly exposed surfaces of concrete broken from the structure or on split cores. All of the fourteen samples changed their color to pink as shown in Figure15. This showed that no concrete carbonation was detected for CWICS up to the present, despite the fact that some of the trains have been in service for more than 30 years.

Figure 15: Carbonation test of core-drilled samples.

Determine Yields Strength of Rebar and Remaining Steel Rebar Thickness

Yield strength of rebar can be obtained from available as-built drawing. However, to get a more accurate data of yield strength, tensile test was performed. Reinforcing bars extracted during concrete core-drilled were used as specimen samples. Four samples of rebar were successfully tensile tested. The result of one of tensile test is shown in Figure 16. The yield strength of rebar was found between 533 and 560 MPa, whereas their corresponding ultimate strengths were found between 759 and 878 MPa. This yield strength was higher than that of the specified yield strength of 400 MPa.

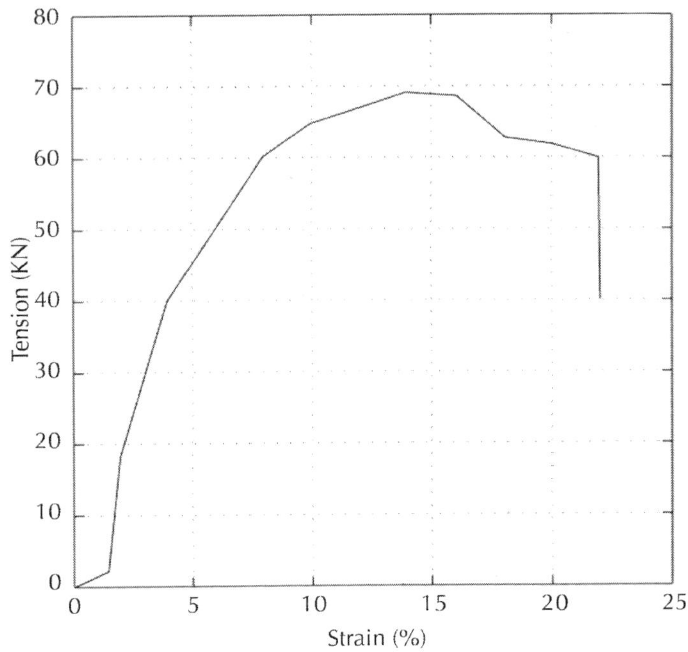

Figure 16: Force and displacement of rebar from tension test.

The loss of rebar thickness due to corrosion was measured using Olympus metallurgical camera and Union metallurgical microscope, as shown in Figures 17 and 18. The rebar samples for this test were obtained from concrete core-drilled. Table 5 shows the corrosion thickness of the rebar for each train. This table shows that train E/F has the highest corrosion rates of 0.1875–0.5 mm/year. This corrosion rate is much higher than the corrosion rate of trains A/B and C/D at 0.0003–0.0006 and 0.0007–0.0017 mm/year, respectively. These two trains have almost negligible corrosion. The higher corrosion rate observed at train E/F is most likely due to local incidence such as local low concrete compaction. Therefore, this value should not be used as a representative value of the steel corrosion rate of train E/F. Furthermore, as only one sample was taken for this test for each train, this result should be used cautiously and should be compared with other forms of tests or formulae to determine the corrosion rate of CWICS. The representative value of corrosion rate of each train will be discussed and determined in Section 4.8.

Table 5: Corrosion thickness of rebar

Number	Code	Rebar diameter	Train	Corrosion thickness	Corrosion rate (mm/ year)	Microstructure
1	Core 3	***D*** 19	A/B	10-20ym	0.0003-0.0006	Pearlite and Ferrite
2	Core 8	***D*** 19	C/D	20-50pm	0.0007-0.0017	Pearlite and Ferrite
3	Core 15	***D*** 19	G/H	3-8mm	0.1875-0.5	Pearlite and Ferrite

Number	Code	Rebar diameter	Train	Corrosion thickness	Corrosion rate (mm/ year)	Microstructure
1	Core 3	19	A/B	10~20 μm	0.0003–0.0006	Pearlite and Ferrite
2	Core 8	19	C/D	20~50 μm	0.0007–0.0017	Pearlite and Ferrite
3	Core 15	19	G/H	3~8 mm	0.1875–0.5	Pearlite and Ferrite

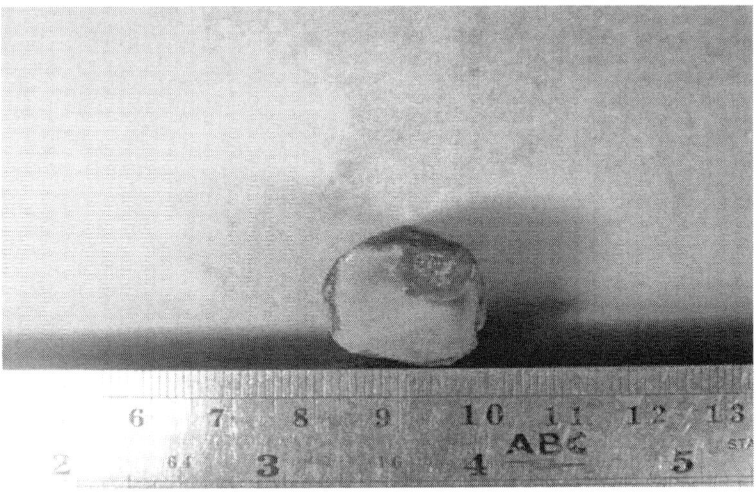

Figure 17: Corrosion thickness of 3 ~ 8 mm taken from core 15

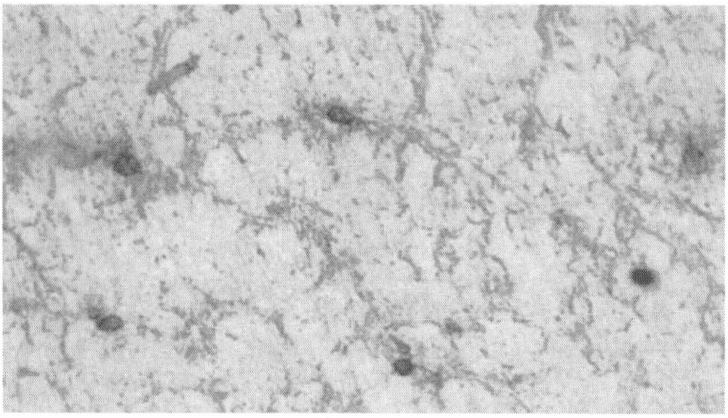

Figure 18: Microstructure with 500x magnification of rebar taken from core 15.

Determine Depth and Chloride Content and Concrete pH in the Concrete

After compression test of core-drilled sample, the debris from this test was chloride tested. Three different depths were used to measure the chloride content, that is, 0.0, 2.5, and 5.0 cm from concrete surface. At the same time, the pH of the concrete was also measured. The results of chloride test and pH test are presented in Figures 19 and 20, respectively.

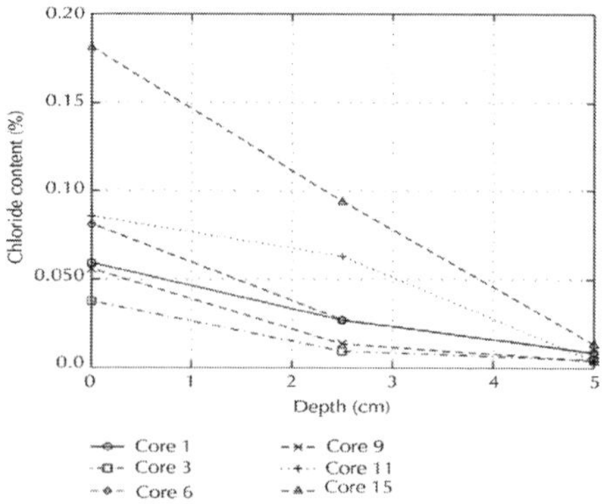

Figure 19: Chloride content at different depths.

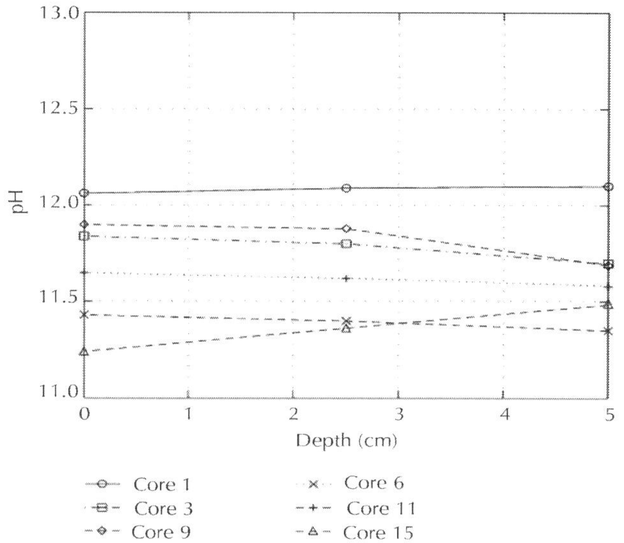

Figure 20: pH value at different concrete depth.

Figure 19 shows that the chloride content measured by weight of concrete (in %) at concrete surface for all samples is very close to one another, except sample taken from core 15. Core 15 drilled at train G/H shows the highest chloride content at all measured depths. This highest chloride content correlates with its highest thickness loss of rebar as presented in Table 5. Figure 19 also indicates that all of the samples have a very similar chloride content of 0.01% at concrete depth of 50 mm, where the rebar is located. This value can be compared with the chloride threshold level to initiate corrosion of 0.025% such as stipulates in Indonesia Concrete Building Code [2].

Figure 20 shows that the concrete pH is relatively constant as the depth from concrete surface increases. The lowest pH is 11.25 at concrete surface and 11.35 at 5 cm depth. This indicates that the concrete is still in a very high alkaline condition and has no experience pH reduction due to corrosion attack. This result corroborates with previous result (see Figure 19) which indicate that concrete corrosion has not yet initiated at CWICS. Note that core 15 which has the highest chloride content also has the lowest pH at concrete depth of 0.0 and 25 mm and the second lowest pH at concrete depth of 50 mm. Core 15 also has the highest thickness loss of rebar as shown in Table 5.

Determine Chloride and Sulphate Content of Sea Water

Sea water surrounding CWICS was tested to determine the concentration of its main aggressive elements that influence the degree of chloride attack. Two samples were tested and the results are presented in Table 6. This table shows that the highest chloride and sulphate content of the sea water is 14250 mg/L and 1600 mg/L, respectively. These values are lower than the chloride and sulphate content of sea water found in the Persian Gulf [13] at 26800 mg/L and 3460 mg/L, respectively. These lower contents are possibly caused by high water rainfall in Indonesia than that in the Persian Gulf.

Table 6: Main aggressive element in sea water.

Parameter	Unit	Sample	
		1	2
pH		7.77	7.82
Sulphate	mg/L	1600	1585
Chloride	mg/L	14250	14240

Determine the Probability of Corrosion of Rebar

The risk of corrosion of rebar in concrete can be estimated using half-cell potential test. Half-cell potential test is simple, cheap, and nondestructive. The electrode used for this test is copper/copper sulphate electrode (CSE). The test was performed based on ASTM [14]. The result of this test is summarized in Table 7.

Table 7: Half-cell potential measurement of rebar

Number		Train A/B	Train C/D	Train E/F	Train G/H
		Readings (mV)			
1		−0.380	−0.120	−0.080	−0.090
2		−0.330	−0.130	−0.090	−0.070
3		−0.220	−0.090	−0.120	−0.100
4		−0.370	−0.030	−0.100	−0.070
5		−0.180	−0.120	−0.110	−0.080
6		−0.220	−0.090	−0.130	−0.100
7		−0.120	−0.130	−0.150	−0.120
8		−0.130	−0.280	−0.130	−0.060
9		−0.180	−0.140	−0.100	−0.080
10		−0.200	−0.140		
11		−0.220	−0.160		
12		−0.150	−0.120		
13		−0.180	−0.240		
14		−0.210	−0.190		
15		−0.140	−0.180		
16		−0.050	−0.200		
17		−0.050	−0.270		
18		−0.050	−0.320		
19		−0.050	−0.270		
20		−0.100	−0.270		
21			−0.270		
22			−0.340		
23			−0.510		
24			−0.520		
25			−0.520		
	Minimum	−0.380	−0.520	−0.150	−0.120
	Maximum	−0.050	−0.030	−0.080	−0.060
	Average	−0.177	−0.226	−0.112	−0.086

Table 7 shows that the most negative potential of rebar (i.e., −0.520 mV) was found at train C/D, followed by trains A/B, E/F, and G/H. All potential readings indicate that the potential of rebar is already in negative side. According to ASTM C-876, the potential reading less than −350 mV means that the probability of corrosion of rebar is greater than 90%. If the result of potential measurement is combined with pH test (i.e., −0.520 mV and pH 11.35) and then plotted using Roubaix diagram, then the corrosion tendency of rebar can be seen in Figure21. This figure shows that the concrete of CWICS is still in no corroding stage (at passivation zone). This result confirms the result of corrosion rate measurement discussed in Section 4.3 which indicates that the train has almost negligible corrosion rate as found in cores 3 and 8. However, this condition may turn in to corroding stage if the concrete pH decreases to less than 10.0.

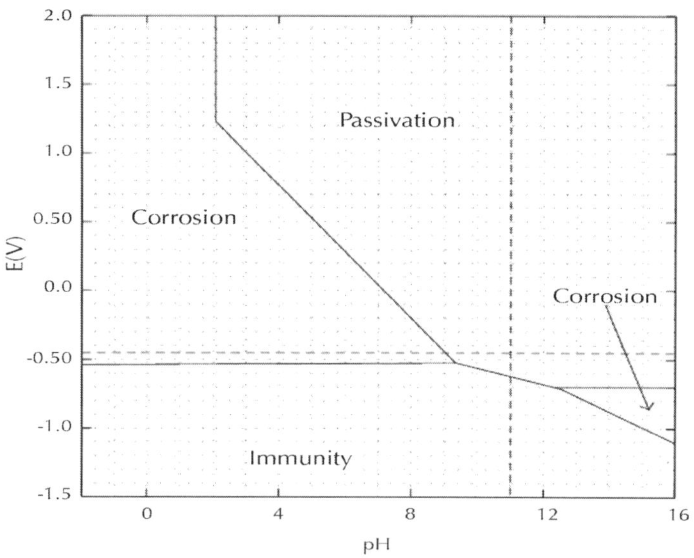

Figure 21: Roubaix diagram of rebar.

Structural and Load Modelling of CWICS

Structural and load modeling of CWICS was performed using SAP 2000 to determine the internal forces of CWICS. These internal forces were then compared with the remaining capacity of CWICS's structural elements. The remaining capacity of CWICS has decreased from its initial design capacity due to rebar corrosion. If ratio of the capacity of concrete element to the internal force of the element (defined herein as safety factor) is greater than unity, the element is considered in a safe condition. However, if this ratio reaches unity or less, the element theoretically has failed and has to be strengthened to achieve the minimum safety of 1.0. Note that the redundancy effect of this highly indeterminate structure was not considered in this analysis when safety factor of element was determined. Therefore, it must be mentioned here that the actual safety factor of CWICS may be higher than the calculated safety factor obtained from this analysis.

Figure 22 shows structural model of CWICS of train A/B. The structure comprises beam, plate (slab), and wall elements. Loads considered in this analysis were dead, live, equipment, and earthquake load. To get the maximum internal forces in the concrete element, different load combination is determined based on Indonesia Concrete Code [2]. The distribution of bending moment of train A/B due to dead and live is shown in Figure23.

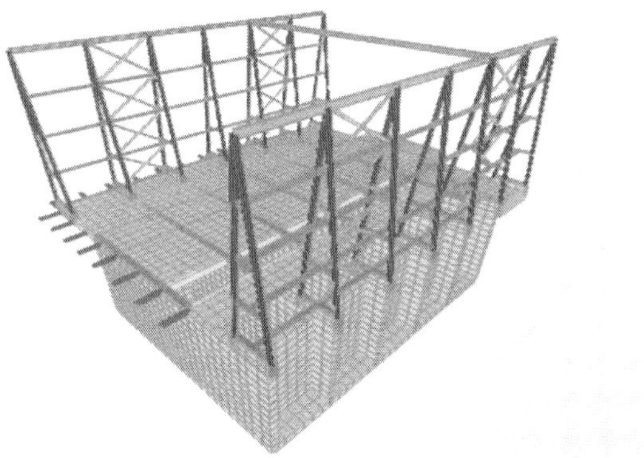

Figure 22: Structural model of train A/B.

Figure 23: Bending moment distribution of train A/B.

Determine Rate of Deterioration

Rate of concrete deterioration or corrosion rate can be determined by two following methods. These are:

- direct method;
- Indirect method.

Direct method of corrosion rate estimation can be performed by measuring either weight loss or thickness loss of rebar. This method requires the steel sample to be extracted from existing structure. Table 5 shows the result of direct method of corrosion rate measurement of CWICS. This table shows that trains A/B and C/D have much lower corrosion rate than the newer train G/H. In this case, this methods yield a contradictory results with actual field condition of CWICS, which shows that older train shows more sign of distress than newer train. For this reason, indirect method of corrosion rate measurement was employed in this study and direct method was mainly used for comparison purpose only.

Indirect method of corrosion rate estimation was performed by using empirical formulae available in many literatures. These formulae were developed through three decades of research on corrosion mechanism and will be discussed briefly in the following paragraph. To estimate concrete rate of deterioration using indirect method, the actual concrete cover needs to be measured. The thickness of concrete cover determines the resistant of concrete structure against corrosive agent such as chloride. To initiate corrosion, chloride must penetrate concrete cover, reach the rebar level, and accumulate to chloride threshold level. In this study, concrete cover was measured using Photometer 5+. The result of this test is summarized in Table 8. Note that based on the available document, the specified concrete cover was 75 mm.

Table 8: -Concrete cover thickness

Location	Concrete cover (mm)	
	Min	Average
Train A/B	50.0	83.3
Train C/D	32.5	65.0
Train E/F	53.0	74.0
Train G/H	44.0	61.7

Table 8 shows that the lowest average concrete cover is 61.7 mm found in train G/H. This value can be compared with the minimum cover thickness as specified in Indonesian Concrete Standard [2], which stipulates that the minimum cover for corrosive environment is 65 mm. However, all the minimum cover found during the test does not comply with the present day code requirement. The actual concrete cover found during this test can be also used as indication of quality control during construction phase. It is very surprising that the oldest train (train A/B) shows a better quality in terms of cover thickness than the newer trains. Train A/B has the highest average concrete cover and the highest minimum concrete cover at 83.3 mm and 50.0 mm, respectively.

Deterioration stage of reinforced concrete structure subjected to corrosion can be divided in two stages [15]:

- Corrosion initiation;
- Corrosion propagation.

The time required for the chloride concentration at the steel surface to reach the threshold chloride concentration needed to destroy the passive layer of the steel is defined as corrosion initiation. The second stage is called corrosion propagation, where steel reinforcing bar corrodes causing loss of area (metal loss) and reduces flexural and shear strength.

Corrosion initiation can be determined using Fick's second law [16] as

$$T_i(C_0, C_{th}, D, d_c) = \frac{d_c^2}{4D\left[erf^{-1}(1-(C_{th}/C_0))\right]^2}, \quad (1)$$

Where C_0 = chloride content at concrete surface, C_{th} = threshold chloride content to initiate corrosion, d = concrete diffusion coefficient, d_c = concrete cover, and erf = the error function.

Chloride content at concrete surface (C_0) has been determined from chloride test discussed at Section 4.4, while threshold chloride content to initiate corrosion (C_{th}) is prescribed in most concrete codes or used empirical values found in the literature. The average and minimum value of concrete cover shown in Table 8 can be used in corrosion initiation calculation to obtain two scenarios of deterioration, that is, average and worst case scenarios.

The concrete diffusion coefficient in (1) can be estimated using empirical formulae [17] as

$$\frac{w}{c} = \frac{27}{f'_{cyl} + 13.5}, \quad (2)$$

Where w/c is the water cement ratio and f'_{cyl} is the cylinder compressive strength from core-drilled concrete.

Corrosion propagation is determined using empirical formula [15] as

$$i_{corr} = \frac{27.0(1-w/c)^{-1.64}}{d_c} (\mu A/cm^2) \quad (3)$$

Where i_{corr} is corrosion rate in μ_A/cm^2. Note that a corrosion current density of 1 μ_A/cm^2 is equal to a steel section loss of 11.6 μ_m/year [18].

The above formulae is used to predict the corrosion rates in concrete structures for mean relative humidity (RH) of 80% and mean temperature of 20°C. To obtain corrosion rate at different temperature, the following formulae [5] can be utilized:

$$i_{corr}(t)=i_{corr-20}[1+0.073\times(t-20)] \tag{4}$$

Where $i_{corr}(t)$ = corrosion rate temperature > 20°C, = corrosion rate temperature 20°C, and = temperature (°C).

At the present study, the average temperature used was 31°C. Using this value and (4), the corrosion rate increases around 80% compared with corrosion rate at 20°C.

Assuming general uniform corrosion, as displayed in Figure 24, the diameter reduction of reinforcing bar (rebar) due to corrosion can be estimated as

$$\Delta D(T_0)=0.0232\times i_{corr}\times T_0\,. \tag{5}$$

The remaining area of rebar can then be determined as

$$A_s(T_0)=\frac{\pi}{4}\left(D_0-0.0232\times i_{corr}T_0\right)^2 \tag{6}$$

Where T_0 is time measured after corrosion initiation.

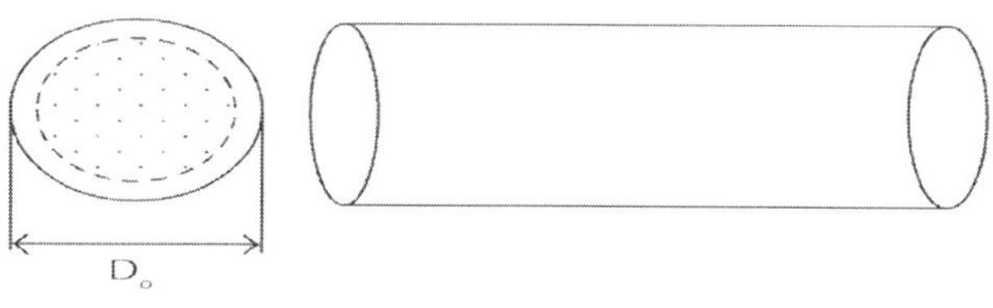

Figure 24: General uniform corrosion.

Using (1) to (6), concrete deterioration then can be determined. Two scenarios were used for this study:

- worst case condition scenario;
- Average case condition scenario.

In the worst case scenario, all parameters used in the analysis were either the minimum or the maximum value obtained from the test in order to get the fastest deterioration of structure. For example, the minimum value was used for concrete cover thickness and concrete strength parameter, whereas the maximum value was used for chloride content parameter. By contrast, average values of parameters were used for the average case scenario. Table 9 summarizes the results of analysis using these two scenarios. Note that concrete strength used in this analysis is concrete strength obtained from compressive test of concrete core-drilled samples.

Table 9: Corrosion initiation time and concrete deterioration rate

Train	A/B		C/D		E/F		G/H	
Scenarios	Worst case	Average case	Worst case	Average case	Worst case	Average case	Worst case	Average case
Built (year)	1977		1982		1987		1995	
Operated (year)	1979		1984		1989		1997	
Compressive strength (MPa)	22.2	24.9	37.8	39.6	35.1	37.4	41.1	41.1
Cover thickness (mm)	50	83	32.5	65	53	74	44	61.7
Corrosion initiation (year)	1.82	13.91	5.73	35.56	10.25	26.13	6.99	26.13
Corrosion rate (mm/year)	0.229	0.100	0.118	0.056	0.081	0.053	0.079	0.056
Remaining capacity at 2013 (%)	44%	83%	77%	100%	90%	100%	94%	100%
Remaining capacity in 20 year (%)	21%	67%	59%	93%	77%	92%	80%	92%

Table 9 shows that train A/B has the shortest corrosion initiation time for the worst and average scenarios as it has the lowest compressive strength. This train also has the highest corrosion rate at around 0.229 mm/year for worst case scenario and 0.1 mm/year for average case scenario, respectively. It is interesting to compare these corrosion rates with corrosion rates obtained using direct method as shown in Table 5. The corrosion rate of train A/B using indirect method for the two scenarios is higher than that obtained from direct method, which gives corrosion rate of 0.0003–0.0006 mm/year. Therefore, corrosion rate based on indirect method yields more conservative results than that from direct method. For this reason, the corrosion rate from indirect method will be used for determining the remaining capacity of CWICS.

Based on the above assumption, for the worst case scenario, the remaining capacity of train A/B at 2013 is 44% of the initial capacity. However, for the average condition scenario, the-2013 capacity of train A/B is around 83% of the initial capacity. This average condition appears to better represent the actual train condition as up to now this train is still in service and there is no indication of significant of distress of the train.

Table 9 also indicates that, for average case scenario, the remaining capacity of train C/D, E/F, and G/H at year 2013 is still 100% of their initial design capacities. By year 2033, these remaining capacities have decreased to 93%, 92%, and 92%, respectively. By comparison, for worst case scenario the remaining capacity of these trains at year 2013 are 77%, 90%, and 94% of their initial design capacities, respectively. By year 2033, these remaining capacities reduce to 59%, 77%, and 80% of their initial design capacity, respectively.

Determine Safety Factor

To better capture the current condition of CWICS, the reduction of safety factor of different element of CWICS due to rebar corrosion against flexure and shear action will be presented. Only the result of analysis of train A/B will be discussed in the next section as this train has the worst condition.

The safety factor of concrete element against flexure and shear can be formulated as

$$SF = \frac{\mu M_n(T_0)}{M_u} \succ 1.0 \tag{7}$$

$$SF = \frac{\mu V_n(T_0)}{V_u} \succ 1.0 \tag{8}$$

Where $M_n(T_0)$ = nominal flexural capacity of concrete element at T_0 after corrosion has initiated, M_u = flexural moment due to factor load obtained from structural analysis, $V_n(T_0)$ = nominal shear capacity of concrete element at T_0 after corrosion has initiated, and V_u = shear due to factor load obtained from structural analysis.

The capacity of concrete element against flexure and shear at T_0 can be determined as

$$M_n(T_0)=A_s(T_0)\times 0.8\times d, \tag{9}$$

$$V_n(T_0)=\frac{1}{6}\sqrt{f_c^{'}\times b_w\times d}+\frac{A_v(T_0)\times f_y\times d}{s} \tag{10}$$

Where A_s (T_0) = area of rebar section at time T_0, F_y = yield strength of rebar, h = height of section, f_c' = concrete compressive strength, b_w = width of section, d = effective depth of section, $A_v(T_0)$ = area of shear reinforcement at time T_0, and s = spacing of shear reinforcement.

The area of rebar for flexure defined as A_s (T_0) and for shear defined as A_v (T_0) can then be determined using (6). For the purpose of this study, the safety was determined at year 2013 and year 2033 using (7) to (10). Figure25 shows the reduction of safety factor for concrete plate element with 600 mm depth and reinforced with 19 mm rebar diameter at 150 mm spacing.

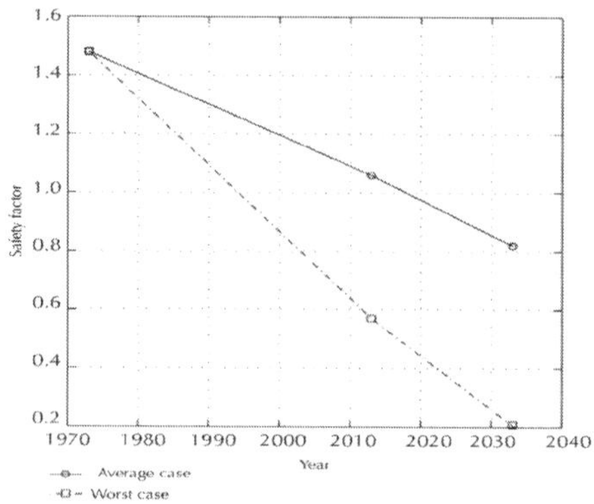

Figure 25: Safety factor for plate element with 600 mm depth and reinforced with 19 mm diameter rebar at 150 mm spacing.

Figure 25 indicates that, for average case scenarios, the safety factor of 600 mm plate element decreases from 1.48 to 1.06 at year 2013 and to 0.82 at year 2033, respectively. By comparison, for worst case scenarios, the safety factor decreases from 1.48 to 0.57 at year 2013 and to 0.21 at year 2033, respectively. Therefore, this element requires immediate strengthening as the safety factor already approaches 1.0 at year 2013.

Figure 26 shows the reduction of safety factor for concrete wall element with 600 mm depth and reinforced with 22 mm diameter rebar at 150 mm spacing. This figure indicates that for average case scenarios, the safety factor of 600 mm wall element decreases from 2.54 to 1.95 at year 2013 and to 1.62 at year 2033, respectively. By comparison for worst case scenarios, the safety factor decreases from 2.54 to 1.20 at year 2013 and to 0.58 at year 2033, respectively. Thus, this element does not require immediate strengthening as the safety factor is still greater than 1.0 at year 2013 for both scenarios.

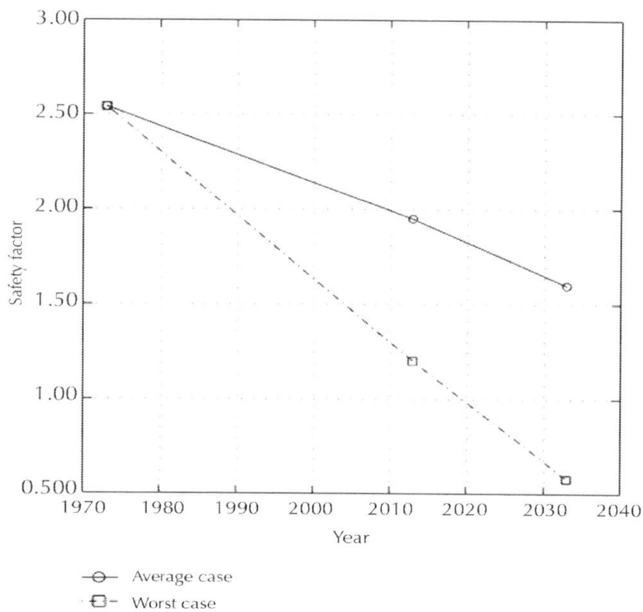

Figure 26: Safety factor for wall element with 600 mm depth and reinforced with 22 mm diameter rebar at 150 mm spacing.

Figure 27 shows the reduction of safety factor for beam element with 500 mm × 800 mm cross-section and reinforced with 4D28 mm diameter rebar against flexure. This figure indicates that, for average case scenarios, the safety factor of the beam decreases from 2.46 to 1.94 at year 2013 and to 1.65 at year 2033, respectively. By comparison, for worst case scenarios, the safety factor decreases from 2.46 to 1.33 at year 2013 and to 0.81 at year 2033, respectively. Thus, this beam does not require immediate strengthening as the safety factor is still greater than 1.0 at year 2013.

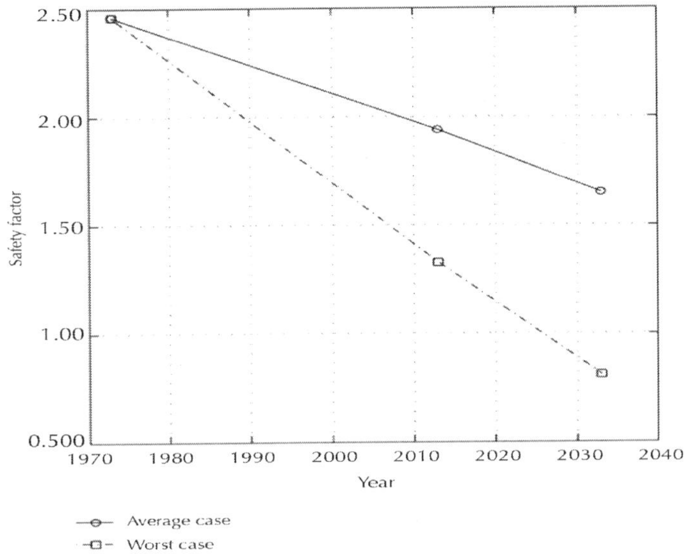

Figure 27: Safety factor for beam element with 500 mm × 800 mm cross-section reinforced with 4D28 mm diameter rebar against flexure.

Figure 28 shows the reduction of safety factor for beam element with 500 mm × 800 mm cross-section and shear reinforced with 2D12 mm diameter rebar with 150 mm spacing against shear. This figure indicates that, for average case scenarios, the safety factor of the beam decreases from 4.36 to 3.39 at year 2013 and to 2.97 at year 2033, respectively. By comparison, for worst case scenarios, the safety factor decreases from 4.36 to 2.48 at year 2013 and to 2.24 at year 2033, respectively. Thus, this beam does not require immediate strengthening as the safety factor is still greater than 1.0 at year 2013.

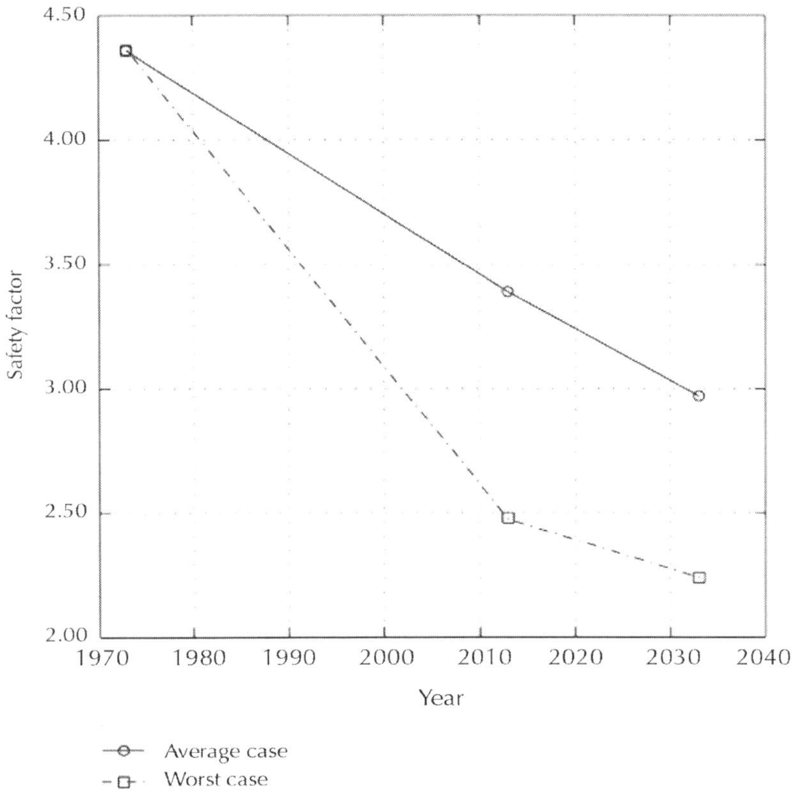

Figure 28: Safety factor for beam element with 500 mm × 800 mm cross-section and shear reinforced with 2D12 mm diameter rebar with 150 mm spacing against shear.

It should be mentioned here that the remaining life assessment of concrete structure due to corrosion attack also has some limitations. Some of the models used in the analysis are derived based on idealized condition. For example, the assumption used for corrosion initiation model based on Fick's second law given in (1) may not be in agreement with the actual service conditions. Fick's second law assumes that concrete is homogeneous material and relative in moist condition (saturated). In reality, concrete cover is generally not saturated with water, concrete is a nonhomogeneous material due to the presence of micro cracking, interconnected pores, and aggregated particles, and the diffusion coefficient may change with time due to hydration progress [19]. Therefore, the remaining life assessment of reinforce concrete structure should be combined with engineering judgment and should be validated with the actual field condition. Further, the remaining life assessment should be performed every 5–10 years as conditions may change significantly than those predicted by available deterioration model.

CONCLUSIONS

The main conclusions drawn from this study can be summarized as follows.

(i) From field and laboratory tests, no significant corrosion activity has been found at CWICS. Most of the reinforcing bars were still in a relatively passive condition as concrete surrounding the reinforcing bars was still in a high alkaline stage. Furthermore, the chloride level at rebar position was found around 0.01% by weight of concrete. This value was still below chloride threshold level to initiate corrosion given in SNI-03-2847 at 0.025%.

(ii) From compressive test of core-drilled sample, train A/B has the lowest average strength of all samples. However, in terms of cover thickness, train A/B has the highest cover thickness of all trains.

(iii) Due to its lowest compressive strength obtained from compression test of core-drilled sample, train A/B has the highest corrosion rate for all trains.

(iv) Based on the available data compiled from the tests, two different scenarios were used to estimate the remaining life of CWICS. Using this approach, the average case scenario represented closer to the actual condition than that of the worst case scenario. The analysis using worst case scenario for train A/B gives the remaining capacity of 40% of the initial capacity. This result does not represent the existing condition of CWICS, which shows no significant sign of distress up to the present. By contrast, using average case scenario for train A/B gives the remaining capacity of 83% of the initial capacity.

(v) Structural analysis shows that the safety factor of most concrete elements of CWICS was still higher than unity up to year 2033. However, fewer elements were also found to have safety factor approach to unity at year 2013. These elements with low safety factor require immediate strengthening to fulfill its intended service life up to 2033.

ACKNOWLEDGMENTS

The authors greatly acknowledge the support of Material Testing Laboratory of Civil Engineering Diploma Program and the Institute of Research and Community Services of Institute Technology Speculum November (ITS), during field and laboratory investigation of the study.

REFERENCES

1. A. Neville, "Chloride attack of reinforced concrete: an overview," Materials and Structures, vol. 28, no. 2, pp. 63–70, 1995. View at Publisher · View at Google Scholar · View at Scopus
2. "Tata Cara Perhitungan Structure Baton untuck Bingaman Gerung (Indonesia concrete building code),"SNI-03-2847, 2013.
3. P. K. Mehta, Concrete in the Marine Environment, Taylor & Francis, New York, NY, USA, 2003.
4. H. S. Müller, M. Heist, and M. Vogel, "Assessment of the sustainability potential of concrete and concrete structures considering their environmental impact, performance and lifetime," Construction and Building Materials, vol. 67, pp. 321–337, 2014. View at Publisher · View at Google Scholar · View at Scopus
5. Dura Crete, Probabilistic Performance Based Durability Design of Concrete Structures, 2000.
6. BS-1881:Part203, Recommendations for Measurement of Velocity of Ultrasonic Pulses in Concrete, British Standards Institution, London, UK, 1986.
7. ACI-Committee-228.1R-03, In-Place Methods to Estimate Concrete Strength Building, American Concrete Institute, Farmington Hills, Mach, USA, 2003.
8. P. K. Mehta and P. J. M. Montero, Concrete: Microstructure, Properties, and Materials, McGraw-Hill, 3rd edition, 2006.

9. RILEM-Recommendations, "Absorption of water by immersion under vacuum. Materials and structures," in RILEM CPC 11.3, vol. 101, pp. 393–394, 1984.
10. X. Chen, S. Wu, and J. Zhou, "Influence of porosity on compressive and tensile strength of cement mortar," Construction and Building Materials, vol. 40, pp. 869–874, 2013. View at Publisher · View at Google Scholar · View at Scopus
11. Y. Y. Kim, K. M. Lee, J. W. Bang, and S. J. Kwon, "Effect of W/C ratio on durability and porosity in cement mortar with constant cement amount," Advances in Materials Science and Engineering, vol. 2014, Article ID 273460, 11 pages, 2014. View at Publisher · View at Google Scholar · View at Scopus
12. BS-EN-14630, Products and Systems for the Protection and Repair of Concrete Structures. Test Methods. Determination of Carbonation Depth in Hardened Concrete by the Phenolphthalein Method, British Standards Institution, London, UK, 2006.
13. M. Shekarchi, F. Moradi-Marani, and F. Pager, "Corrosion damage of a reinforced concrete jetty structure in the Persian Gulf: a case study," Structure and Infrastructure Engineering: Maintenance, Management, Life-Cycle Design and Performance, vol. 7, no. 9, pp. 701–713, 2011. View at Publisher ·View at Google Scholar
14. ASTM-C-876, "Standard test method for half-cell potentials of uncoated reinforcing steel in concrete," in Annual Book of ASTM Standards, vol. 03.02, pp. 11–16, 2006.
15. K. A. T. Vu and M. G. Stewart, "Structural reliability of concrete bridges including improved chloride-induced corrosion models," Structural Safety, vol. 22, no. 4, pp. 313–333, 2000. View at Publisher · View at Google Scholar · View at Scopus
16. J. Zhang and Z. Louis, "Sensitivity analysis of simplified diffusion-based corrosion initiation model of concrete structures exposed to chlorides," Cement and Concrete Research, vol. 36, no. 7, pp. 1312–1323, 2006. View at Publisher · View at Google Scholar · View at Scopus
17. M. G. Stewart and D. V. Godowsky, "Structural safety and serviceability of concrete bridges subject to corrosion," Journal of Infrastructure Systems, vol. 4, no. 4, pp. 146–155, 1998. View at Publisher · View at Google Scholar · View at Scopus

18. D. A. Jones, "Localized surface plasticity during stress corrosion cracking," Corrosion, vol. 52, no. 5, pp. 356–362, 1996. View at Publisher · View at Google Scholar · View at Scopus

19. J.-K. Kim, C.-Y. Kim, S.-T. Yi, and Y. Lee, "Effect of carbonation on the rebound number and compressive strength of concrete," Cement and Concrete Composites, vol. 31, no. 2, pp. 139–144, 2009. View at Publisher · View at Google Scholar · View at Scopus

Chapter 7

Probabilistic Failure Modelling of Reinforced Concrete Structures Subjected To Chloride Penetration

Caio Gorla Nogueira, Edson Denner Leonel* and Humberto Breves Coda

Department of Structural Engineering, School of Engineering of São Carlos, University of São Paulo, Av. Trabalhador SaoCarlense, 400, São Carlos, São Paulo 13566-590, Brazil

ABSTRACT

Structural durability is an important criterion that must be evaluated for every type of structure. Concerning reinforced concrete members, chloride diffusion process is widely used to evaluate durability, especially when these structures are constructed in aggressive atmospheres. The chloride ingress triggers the corrosion of reinforcements; therefore, by modelling this phenomenon, the corrosion process can be better evaluated as well as the structural durability. The corrosion begins when a threshold level of chloride concentration is reached at the steel bars of reinforcements. Despite the robustness of several models proposed in literature, deterministic approaches fail to predict accurately the corrosion time initiation due the inherent randomness observed in this process. In this regard, structural durability can be more realistically represented using probabilistic approaches. This paper addresses the analyses of probabilistic corrosion time initiation in reinforced concrete structures exposed to chloride penetration. The chloride penetration is modelled using the Fick's diffusion law. This law simulates the chloride diffusion process considering time-dependent effects. The probability of failure is calculated using Monte Carlo simulation and the first order reliability method, with a direct coupling approach. Some examples are considered in order to study these phenomena. Moreover, a simplified method is proposed to determine optimal values for concrete cover.

INTRODUCTION

Concrete structures, especially reinforced concrete structures, are one of the most used types of structures around world. When it is located in non-aggressive environments, these types of structures respect, in general, its structural life predicted. However, the structural durability can be strongly reduced by degradation processes of environmental and/or functional origins (Angst ET al.2009; Guzman et al. 2011). Among these processes, it is worth to mention the chloride penetration, carbonation, fatigue, and creep. Durability can be measured as a period of time in which the structure maintains, at least, minimal functional conditions, resistance, and external aspects required in its design.

Considering reinforced concrete structures, durability and service life should be parameters with major importance on rational design. In these cases, analyses and criteria based on environmental aspects, prediction of repairs, and maintenance costs should be required on design stage (Nogueira et al. 2012). These types of analyses are accurately performed considering probabilistic approaches, due the large randomness contained in environmental phenomena. In this regard, the requirement of time-variant reliability assessment of structural durability has becoming increasingly accepted (Xiang and Zhao 2007; Sarveswaran and Roberts 1999; Kong et al. 2002).

Reinforcement's corrosion is directly responsible by durability and failure of reinforced concrete structures. Moreover, chloride ingress is identified as one of the major factors that cause reinforcement's corrosion. When the corrosion begins, the structure is affected by several phenomena like reduction of reinforcement's cross-section, bursting of concrete, and adherence loss between steel and concrete (Kirkpatrick et al. 2002; Jung et al. 2003), all of them reduce the structural durability.

The deterioration process of reinforced concrete structures due to steel's corrosion can be divided into two stages: the initiation and the propagation periods. During the initiation period, chlorides diffuse through concrete toward the reinforcements. The chloride concentration along the concrete cover grows until reaches a threshold value, which triggers the corrosion of steel. The propagation period includes the period of time from the start of corrosion to a critical steel loss. Compared with the first stage, the propagation period is relatively short. Therefore, the chloride diffusion process is frequently used to indicate the durability and service life conditions of concrete structures (Tuutti 1982).

Several models have been proposed in literature to deal properly with the corrosion phenomenon in reinforced concrete structures. Most of them propose deterministic approaches in order to model this structural problem. However, due to the large number of uncertainties, which are inherent at this problem, it can only be accurately analyzed considering probabilistic approaches. Among these models and formulations, it is worth to mention the research of Thirumalai et al. (2006); Enright and Frangopol (1998), DuraCrete (2000), Vu and Stewart (2000), and Balaji Rao et al. (2004), where the corrosion process was studied, considering several conditions and parameters. Based on these probabilistic models, durability and service life assessment analyses in reinforced concrete structures exposed to chloride penetration can be accurately performed (Stewart and Rosowsky 1998; Estes and Frangopol 2003; Suo and Stewart 2009). Moreover, considering its results, maintenance and inspection plans, as well as the concrete cover, can be designed taking into account the randomness of each variable and the cost involved on each procedure.

The penetration of chlorides in concrete is controlled by complex interactions among physical and chemical mechanisms, where several sources of uncertainties are observed (Nogueira ET al.2012). This phenomenon is often simplified, without significant representation loss, by a process governed only by diffusion process. In this regard, these problems can be modelled using Fick's diffusion law (Tuutti 1982; Crank 1975). This classical approach of diffusion considers the apparent coefficient of diffusion as constant along time and homogeneous in space. Moreover, this law assumes the concentration of chlorides on the environment as constant and admits the concrete in saturated condition. By coupling Fick's diffusion law with reliability algorithms, probabilistic analyses of chloride ingress in concrete structures can be performed.

In this paper, a coupled model based on mechanical behavior simulation and reliability algorithms is developed in order to allow probabilistic analyses of reinforced concrete structures subjected to chloride ingress. This model aims at quantifying the probability of corrosion start in reinforced concrete structures based on reliability algorithms. The mechanical model is based on Fick's diffusion law, which is capable to simulate the chloride penetration process in porous materials, as well as simulate the chloride concentration growth along time. Therefore, the chloride concentration at a given cover depth and time can be evaluated with this model. The probabilistic model is developed based on reliability algorithms. In order to determine the probability of failure, two reliability algorithms are considered: first order reliability method (FORM), by coupling directly the mechanical model with FORM, and Monte Carlo simulation. Both algorithms determine the probability of failure considering the failure scenarios achieved by the mechanical model based on Fick's diffusion law.

The coupled model developed is applied to the analysis of probabilistic chloride penetration problem in reinforced concrete structures. Some specific scenarios are considered, which differ by the environmental aggressiveness and concrete properties, and the results obtained are discussed. Considering the developed model, a simplified method is proposed in order to obtain optimal values for concrete cover and time intervals for periodic inspection procedures. This simplified method is the main contribution of this paper.

METHODS

The Mechanical Model

Corrosion of reinforcements induced by chlorides occurs in the presence of oxygen and moisture when the chloride build-up within the structure exceeds a threshold value. Even for carefully constructed concrete structures, with negligible or practically non-chloride inherited at the construction stage, the gradual buildup of chloride content takes place slowly through ingress of chlorides from external sources.

The transport phenomenon associated with the movement of chlorides along structures exposed to aggressive environments is attributed, in most part, to diffusion of chloride ions into concrete pores under a concentration gradient. The coefficient of chloride diffusion, which depends upon the pore structure of concrete, characterizes this flow under a given external concentration of chloride. This parameter is assumed as a characteristic property of hardened concrete.

In order to simulate the chloride ingress and its transport into concrete pores, Fick's diffusion law (Crank 1975) has been widely applied and considered as an acceptable model. Fick's laws for diffusion are applicable for homogeneous, isotropic, and inert materials (Dhir et al. 1998). Moreover, the mechanical properties related to diffusion process are assumed to be identical along all directions and kept constants along time. Considering concrete, these hypotheses are not completely satisfied, because concrete is well known as heterogeneous, anisotropic, and chemically reactive (continued hydration and micro-cracking process) material. However, the methods commonly adopted for chloride transportation modelling in concrete consider this process as governed only by ionic diffusion. Then, it assumes that the concrete cover is completely saturated. Therefore, it makes the hypotheses of Fick's laws acceptable for chloride ingress modelling, because, in this case, the material is assumed completely saturated, with unidirectional chloride flux, i.e., from the exterior surface into the concrete depth. When chloride diffuses into porous concrete, a change in chloride concentration, C, occurs at any time, t, in every point, x, of the concrete structure, i.e., it is a non-steady state of diffusion. In order to simplify its analysis, the diffusion problem is considered as one-dimensional.

Many engineering problems of chloride ingress, as those discussed in this paper, can be solved considering this simplification.

The assumption of Fick's diffusion theory is that the transport of chlorides into concrete though a unit section area of concrete per unit of time (the flux F) is proportional to the gradient concentration of chlorides measured at normal direction of section. Then:

$$F = -D\frac{\partial C}{\partial x} \qquad (1)$$

The negative sign on the equation above arises because the diffusion of chlorides occurs in the opposite direction of the increasing concentration of chlorides. The constant of proportionality D, presented in Equation 1, is called chloride diffusion coefficient. In general, D is not constant but depends on many parameters as the time for which diffusion has taken place, location in the concrete, and composition of the concrete among other factors. If the chloride diffusion coefficient is constant, Equation 1 is usually referred as Fick's first diffusion law. If this is not the case, the relation is usually referred as Fick's first general diffusion law.

There are some cases where this simple relation should not be applied. In this regard, it is worth to mention the cases where the diffusion process may be irreversible or has a history dependence. In such cases, Fick's diffusion law is not valid, and the diffusion process is referred as anomalous. However, non-observation so far indicates that the chloride diffusion into concrete should be characterized as an anomalous diffusion. Fick's second law can be derived considering the mass balance principle. Therefore:

$$\frac{\partial C}{\partial t} = \frac{\partial}{\partial x}\left(-D\frac{\partial C}{\partial x}\right) \qquad (2)$$

In order to apply Fick's second diffusion law, in this form, for concrete exposed to chloride during a long period of time, one ought to know the variation of the chloride diffusion coefficient along time t. If only few observations exist in a specific case, it is possible to estimate upper and lower boundaries for the variation of D in time. Despite this dependence, an especial case can be considered where the chloride diffusion coefficient is independent of location, x, time, t, and chloride concentration, C. In this case, Fick's second law can be written in the simple form:

$$\frac{\partial C}{\partial t} = -D_0 \frac{\partial^2 C}{\partial x^2} \quad (3)$$

In which D_0 is the constant coefficient of diffusion.

The solution of the differential equation presented above, for a semi-infinite domain with a uniform concentration at the structural surface, is given by:

$$C(x,t) = C_0 erfc\left[\frac{x}{2\sqrt{D_0 t}}\right] \quad ,(4)$$

Where C_0 is the chloride concentration at the structural surface supposed constant in the time;erfc is the complementary error function.

In this paper, Equation 4 is used to evaluate the chloride concentration, C(x, t), at a given depth and time into reinforced concrete structure. Therefore, Equation 4 is an analytical solution which composes the mechanical model. Based on the chloride concentration values at a given structural depth, it is possible to assess the structural safety. However, in order to allow the safety assessment, a coupling among the described mechanical model and reliability algorithms has to be considered. These coupling models take into account the randomness that emerges on the diffusion process.

Based on these models, the structural safety variation along time can be assessed. Therefore, one of the most important products of these coupling is the possibility to choose the time of structural maintenance based on a given reliability index target. Moreover, these models allow the determination of the cover depth, considering a structural safety level target. The determination of the cover depth and time for maintenance considering uncertainties related to mechanical structural behavior is the main contribution of this paper.

Reliability Concepts and Methods of Analysis

General Concepts

The reliability analyses aim at computing the probability of failure regarding a specific failure scenario known as limit state. The first step in the reliability assessment is to identify the basic set of random variables $X = [x_1, x_2..., x_n]^T$ from which uncertainties have to be considered. For all these variables, probability distributions are attributed in order to model its randomness. These probability distributions can be defined by physical observations, statistical studies, laboratory analysis, and expert's opinion. The number of random variables is an important parameter to determine the computing time consumed during the reliability analysis. In order to reduce the size of the random variable space, it is strongly recommended to consider all variables as deterministic, the uncertainties of which lead to minor effects on the value of probability of failure.

The second step consists of defining a number of potentially critical failure modes. For each of them, a limit state function, G(X), separates the space into two regions as described in Figure 1: the safe domain where $G(X) > 0$, and the failure domain where $G(X) < 0$. The boundary between these two domains is defined by $G(X) = 0$, known as the limit state itself. It is worth to mention that an explicit expression of the limit state function is usually not possible. When numerical mechanical models are used, it can be computed only at a desired number of points. In this paper, the limit state is defined using the critical failure modes calculated using Equation 4.

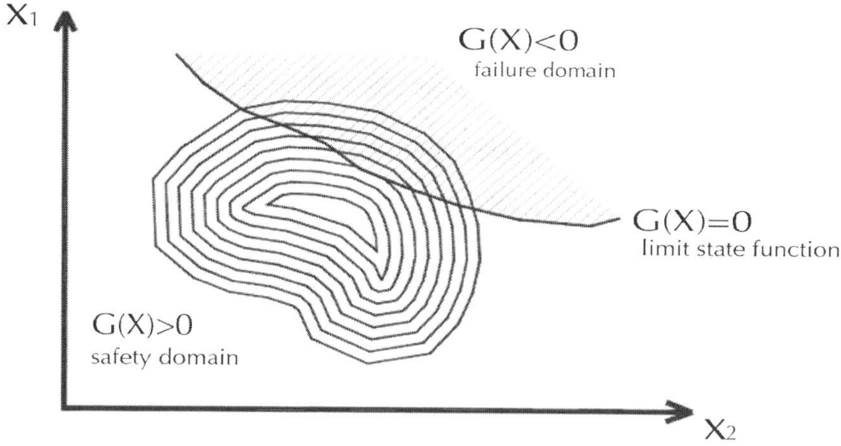

Figure 1: Domains of failure and safety for two random variables.

The probability of failure is evaluated by integrating, over the failure domain, the joint density function as presented by (Ditlevsen and Madsen 1996):

$$P_f = \int_{G \leq 0} f_X(x1, x2, ...Xn)dx_1, dx_2, dx_n, \quad ,(5)$$

In which $f_X(X)$ is the joint density function of the variables, X.

As the evaluation of the integral defined by Equation 5 is almost impossible in practice, alternative procedures have been developed on the basis of reliability index concept, , (Hasofer and Lind1974). This parameter is defined by the distance between the mean point and the failure point placed at the limit state function G(X) = 0 in the normalized space of random variables. The reliability index allows us to calculate the probability of failure using the FORM as follows:

$$P_f = (-), \quad (6)$$

in which (·) is the standard Gaussian cumulated distribution function.

There are some alternative procedures, available in the reliability theory, which allow the evaluation of probabilities of structural failure.

These procedures are based on numerical simulation techniques. The most important approach, among them, is the Monte Carlo simulation method. However, when numerical mechanical models (expensive in terms of computational work) are adopted, this approach may be unreliable due to the large sampling required for simulation, especially when the probability of failure to be assessed is small, i.e., 10^{-3} or lesser.

In this study, two reliability approaches are adopted. The first is known as direct method, as it is the result of the direct coupling of a mechanical model at the FORM. The second approach used is the classical Monte Carlo simulation. In both cases, the models are used to evaluate the probabilities of failure considering the chloride ingress process in reinforced concrete structures. These approaches will be discussed in the following sections.

Direct Method

The basic procedure in this approach consists of directly coupling the reliability model FORM with the mechanical model given by Equation 4. This approach has demonstrated to be accurate and robust for analysis of many complex engineering problems, as discussed by Leonel et al. (2010, 2011a, 2012).

As described in the previous section, the limit state function defines the boundary between safe and failure domains. Considering chloride penetration problem, the limit state function can be written in terms of time for corrosion initiation:

$$G(X) = t_R(X) - t_a, \qquad (7)$$

In which t_R is the time for corrosion initiation that depends on the group of random variables X; t_a is the structural life-time expected in design which was considered as a deterministic parameter.

The time TR is evaluated using Equation 4 assuming the chloride concentration C(x, t) as known at a given position x inside the concrete. Actually, C(x, t) is assumed to be equal to the chloride concentration threshold over which the steel corrosion is triggered on failure condition. The range of x position, in this case, is the concrete cover. It assumes zero at the external surface of the structural member and the cover value at the reinforcement's surface inside the concrete. In this regard, the time for corrosion initiation can be determined explicitly from Equation 4 as:

$$t_R = \frac{1}{D_0}\left\{\frac{z}{2\mathrm{erfc}\left[C(x,t)/C_0\right]}\right\}^2 . \tag{8}$$

In order to include invariance measure of safety, the random variables, defined on physical space, are transformed into independent standard Gaussian variables (Hasofer and Lind 1974) using appropriate probabilistic transformations. The limit state function, defined on physical space, G(X) = 0, is transformed into H (U) = 0 on standard normalized space, where $U = u_1, u_2 ..., u_n^T$ denotes the standard Gaussian variables.

In this standard space, the reliability index is given by the minimum distance between the failure domain and the origin of the standard space. The reliability index can be evaluated by solving a constrained optimization problem, as described below:

Find P*

Which minimizes

$$\beta = \sqrt{U^T.U} \tag{9}$$

Subject to H (U) = 0.

The solution of this problem converges to the failure point nearest to the space origin, known as design point or the most probable failure point P*. In standard space, the distance between this point and the origin is known as reliability index, as shown in Figure 2. The reliability index can be achieved by applying any optimization algorithm. A particular algorithm, which is efficient in this case, to solve reliability problems is the HLRF algorithm (Rackwitz and Fiessler 1978). This optimization algorithm can be coupled directly to the mechanical model. As the time for corrosion initiation is known point-by-point, the resistance t_R is known. Consequently, the limit state function is determined point-by-point. Then, the gradient of the limit state function can be determined using any numerical procedure. In this paper, the forward finite differences technique for this proposal was adopted.

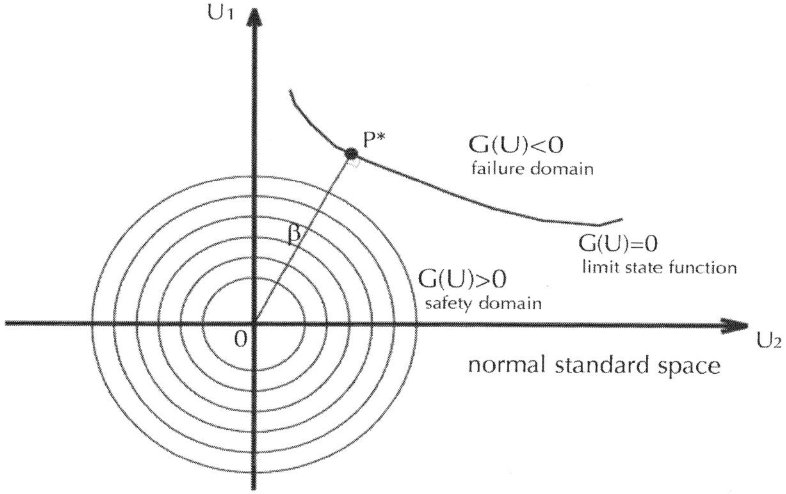

Figure 2: Reliability index and design point.

Some difficulties arise from Equation 8, particularly its gradients evaluation, due to the presence of the complementary error function. A natural barrier that can be remarked in this type of approach is the numerical error due to the finite difference procedure, which may affect the convergence of the coupled procedure, as well as the precision of the solution, especially for nonlinear transient phenomena. However, for all problems studied in this paper, numerical problems related to finite differences method were not observed. Moreover, it was verified that the direct coupling procedure gives accurate results and stable convergence rate with low number of mechanical analyses.

Monte Carlo Simulation

Monte Carlo method is a numerical simulation approach widely used in reliability problems (Nowak and Collins 2000). In this method, a sampling of random variables is used to construct a set of values aiming to describe the failure and safe spaces, and calculate Equation 5. The sampling is constructed based on the statistical distributions assigned for each random variable introduced in the problem. As this method deals with simulations of the limit state function, the bigger the sampling adopted, the more accurate will be the description of spaces and the probability of failure achieved.

The kernel of this method is the construction of a sampling for each random variable involved in the problem. Then, the domain of safety and failure points are prospected by simulating Equation 7, as described in Figure 3. The probability of failure is calculated, for Monte Carlo simulation, using the following expression:

$$P_f = \int_{G \le 0} f_x(x_1)dx_1 = \int_{G \le 0} I(x_1)f_x(x_1)dx_1 = E[I(x_1)]. \quad (10)$$

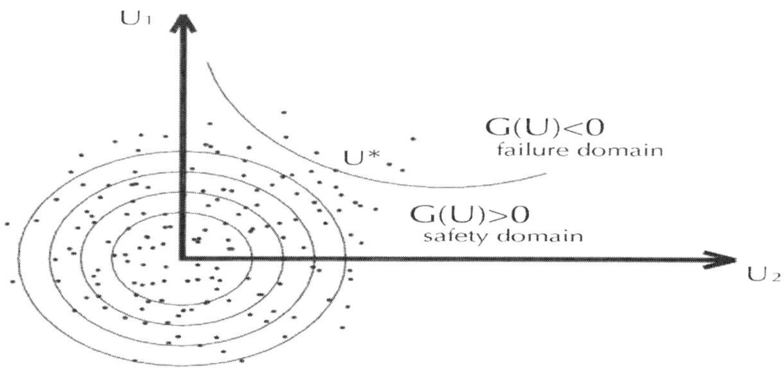

Figure 3: Monte Carlo's sampling for two random variables.

The function I (x_i) can be written as:

$$I(x_1) = \begin{cases} 1 \rightarrow G \le 0 \\ 0 \rightarrow G > 0 \end{cases} \quad (11)$$

By simulating the limit state function for a convenient range of sampling, the mean value of I (x_i) will be an estimator for the probability of failure. Then:

$$\overline{P_f} = E[I(x_1] = \frac{1}{N}\sum_{t=1}^{N} I(x_1). \quad (12)$$

The disadvantage of this method is related to the large amount of simulations required to compute accurately the probability of failure. Normally, in order to estimate accurately the probability of failure of 10^{-n}, the number of simulations must be higher than 10^{n+2} or 10^{n+3}. It means, in civil engineering structures, where the probabilities of failure are in between 10^{-3} to 10^{-6}, at least 10^5 to 10^9 realizations of the limit state function is required. When complex numerical mechanical models are involved, which lead to high computational work, this method may be not reliable. However, theoretically, this method leads to the real probability of failure when the sampling range becomes infinite.

Methodology of Analysis

The corrosion phenomenon modelling in reinforced concrete structures has to take into account two different stages, as illustrated in Figure 4. The first one is related to chloride ingress into porous concrete. In this stage, the chloride concentration, along the cover depth, increases as the time passes. Then, the corrosion starts when the chloride concentration surrounding the reinforcements reaches a threshold level, leading to the loss of the concrete chemical passive protection. At the end of this stage, the steel of reinforcements remains undamaged. The second stage, called propagation stage, is characterized by the reduction of the reinforcement's cross-sections, which generates the loss of structural strength along the time.

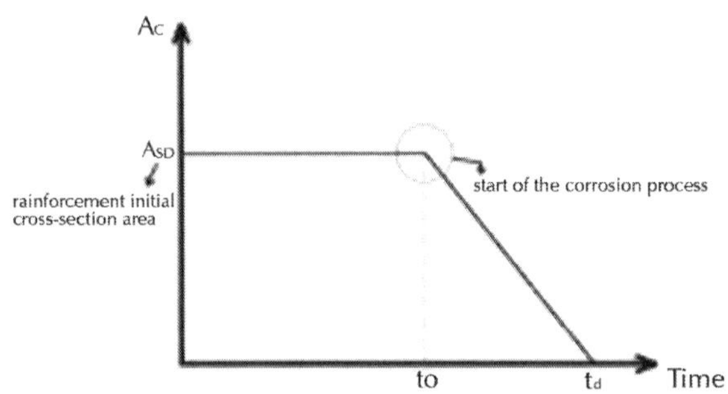

Figure 4: Corrosion process of the reinforcement cross-section area.

Compared with the first stage, the propagation period is relatively short. Therefore, the time for corrosion initiation has been widely adopted for structural durability and safety assessments (Nogueira et al. 2012). In this regard, the objective of this paper is to assess the probability of failure considering the failure scenario predicted by initiation stage. Therefore, the failure is observed when the chloride concentration at the reinforcement's depth reaches the threshold level. The main parameters considered for all reliability analyses performed are the following:

- Chloride concentration threshold at the interface between concrete and reinforcement bars, which defines the beginning of the corrosion process, C(x, t). This parameter was studied experimentally by Vu and Stewart (2000).
- Chloride concentration at the structural surface, C_0. This parameter is related to the environment aggressiveness, and its reference's values may be determined by experimental observations, as presented by Vu and Stewart (2000), or defined using an international concrete standard design (Brazilian Association of Technical Standards 2003).
- Concrete diffusion coefficient, D_0, which has been studied by Papadakis et al. (1996).
- Structural depth, which in this study, is defined as the reinforcement concrete cover, x.

It is worth to stress that initial cracks due concrete cure and/or bending/shear effects and longitudinal cracking have not been considered in the formulation presented in this paper. These phenomena affect the corrosion process, and its modelling can be accurately performed using numerical methods as finite element method and boundary element method (Leonel and Venturini 2011; Leonel et al. 2011b).

According to V_u and Stewart (2000), the chloride concentration at structural surfaces is a function of the atmosphere (environment) where these elements are located. The cover depth is also defined according to the environment, which is stated by international concrete standard designs in categories of aggressiveness. In particular, the values shown in Brazilian Association of Technical Standards (2003) have been used in this paper. The coefficient of diffusion of concrete, which represents the material resistance against chloride ingress, depends on the water/cement ratio (w/c). If the proportion of water is high, the empty volumes inside the matrix due to the cure process of the concrete will also be high. Consequently, the material permeability will be high, and the material resistance against the chloride penetration will be low. Therefore, in regions close to the coast, it is strongly recommended to construct reinforced concrete structures with lower water/cement ratio and/or large covers.

The proposed probabilistic model allows the evaluation of the probability of structural failure, taking into account the random variables previously presented. Moreover, this model is capable to describe the dependency relation between the probability of failure and time. Then, the proposed model can be used to solve an interesting structural problem which relates the determination of time intervals for structural maintenance plans based on structural safety.

In order to apply the proposed model on this problem, a given safety level, say target, must be defined. The mentioned safety level may be determined using the standard of JCSS (2001), where the prevention against structural failures is measured based on a target reliability index. As the chloride concentration at the structural cover increases along time, the safety against this failure mode reduces along time. Therefore, the intervals of time for periodic structural maintenance procedures, based on structural safety, are achieved when the reliability index calculated using the proposed model reaches the target reliability index stated by analyst. It is worth to mention that the repair procedures are assumed as perfect, i.e., after the maintenance the structure recovers its initial integrity conditions without chlorides. These intervals are determined as long as the parameters related to material, cover depth, and environment aggressiveness can be defined a priori.

Another application of the proposed model relates to the definition of values for cover depth and concrete properties (w/c ratio) based on a given safety level and the expected structural lifetime. When the intervals of time for structural maintenance and/or the expected structural lifetime are stated a priori, the w/c ratio and the cover depth values can be obtained by simulating the proposed model in order to define the couple of these values which lead the structure to maintain a safety level at least equal to the target reliability index during the specified period of time. In this application, the values of w/c ratio and cover depth are achieved when the reliability index given by the proposed model is equal to the target reliability index.

It is important to emphasize that these procedures do not take into account the costs involved neither in the maintenance procedure nor in the concrete production. However, these applications of the proposed model can be performed if analysts are interested in design considering, exclusively, safety criterion.

RESULTS AND DISCUSSION

In order to illustrate the applicability of the proposed model, probabilistic analyses were performed considering different scenarios and material properties aiming to evaluate the influence of the environment aggressiveness, w/c ratio and cover depth on the probability of structural failure due to chloride penetration. Four applications are considered, and in each of them, some specific aspects of the problem are explored.

Example 1

In this example, the probabilistic model is applied to the assessment of uncertainties related to chloride ingress in reinforced concrete structures. The results were obtained using direct method and Monte Carlo simulation approaches. Considering the direct method, the tolerance for convergence was verified in terms of reliability index and design point coordinate change, which have to be smaller than 10^{-4}. It is worth to mention that considering this approach, only 65 limit state function runs were required in order to achieve the convergence. Moreover, this number of mechanical model runs is the highest observed among all scenarios considered. It confirms the good performance of direct method approach, which is capable to achieve the convergence with low computational work even in reliability problems involving non-Gaussian random variables and non-linear mechanical model. The analyses performed using Monte Carlo simulation considered a sampling of 10^5 values for each random variable. It leads to 10^5 simulations of the limit state function. Table 1 presents the statistical parameters adopted for all random variables considered in this application, which are based on the work of Vu and Stewart (2000) and of Suo and Stewart (2009).

Table 1: Statistical data for all random variables

Random variable	Mean	Coefficient of variation (%)	Statistical distribution
$C(x,t)$	0.9 kg/m^3	19	Uniform (0.6 to 1.2)
C_0	C.A. II - 1.15 kg/m^3	50	Log-normal
	C.A. III - 2.95 kg/m^3		
D_0	w/c = 0.4 to 14.2 mm^2/year	75	Log-normal
	w/c = 0.5 to 41.0 mm^2/year		
	w/c = 0.6 to 86.4 mm^2/year		
	w/c = 0.7 to 162.7 mm^2/year		
x	C.A. II - 30.0 mm	50	Normal
	C.A. III - 40.0 mm		

Nogueira et al.

According to the Brazilian Association of Technical Standards (2003), categories of aggressiveness (C.A.) II (moderate) and III (high) represent urban regions and coast/industrial areas, respectively. For each one of these C.A., there is a recommended concrete cover depth, x, which was adopted for the probabilistic analyses of each category.

The evolutions of the probability of corrosion initiation along the time, for both categories of aggressiveness considered in this application, are presented in Figures 5 and 6. The results illustrate a considerable growth of the probability of corrosion initiation as time increases. Moreover, a strong dependency between the probability of corrosion initiation and the w/c ratio was also verified. This dependency was expected because the higher the w/c ratio, the higher the concrete permeability and, consequently, easier the chloride ingress will be.

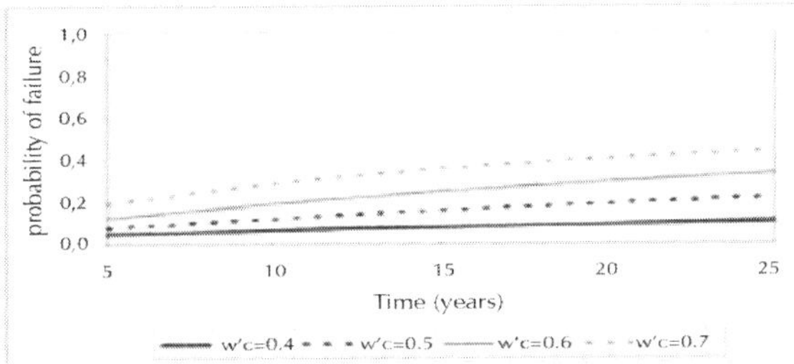

Figure 5: Probability of failure versus time for C.A. II.

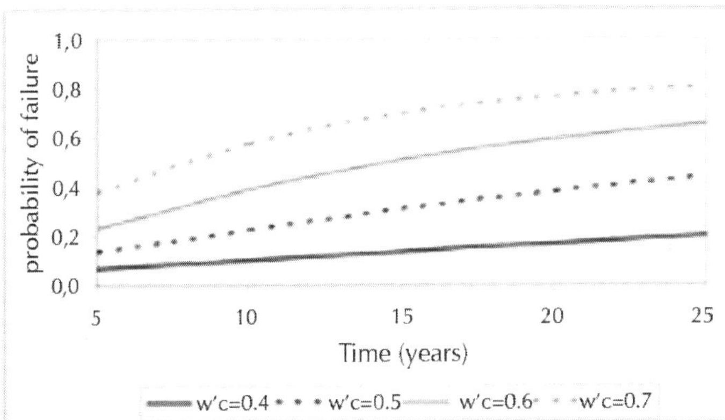

Figure 6: Probability of failure versus time for C.A. III.

Considering the horizon of time analyzed, for concretes with w/c ratio values higher than 0.5, the probability of failure, after 15 years of structural service life, tends to increase slower than in the first 15 years. Therefore, the derivative of probability of corrosion initiation after 15 years is smaller than in the first 15 years. This behavior reflects the chloride concentration evolution along time, which tends to saturate after 15 years. It is important to stress that this behavior was observed in both C.A. analyzed.

Example 2

This application aims at studying the influence of cover depth values on the probability of corrosion initiation considering two different scenarios of environment aggressiveness. The analyses were performed considering concretes composed of w/c ratios varying from 0.40 to 0.70. The analyses involved cover depth in the range of 10 to 60 mm. For each value of w/c, the mean value of cover depth was also varied, but its coefficient of variation was kept constant. This procedure was adopted because this study aims at describing the dependency of probability of failure with respect to w/c and cover depth parameters. The time was assumed as deterministic in this analysis. In order to consider all the combinations, 240 numerical reliability analyses were carried out. The direct method and Monte Carlo simulation were adopted for the evaluation of probability of failure and reliability indexes. Considering the direct method, the convergence is assumed when the difference, into two successive iterations, in terms of reliability index and design point coordinates is smaller than 10^{-4}. Considering Monte Carlo simulation, a sampling of 10^5 values for each random variable was adopted.

The computational time consumed in this analysis is relatively lower. Considering the direct method, the most expensive case required 65 limit state function calls, corresponding to less than 1 s of computational work. The Monte Carlo analyses required 10^5 limit state function evaluation due to the sampling range adopted. However, it took less than 3 s of computational time process. The values adopted for all random variables considered in this application, as well its statistical information, are presented in Table 2, which are based on the work of Vu and Stewart (2000) and of Suo and Stewart (2009).

Table 2: Statistical data for all random variables

Random variable	**Mean**	**Coefficient of variation (%)**	**Statistical distribution**
$C(x,t)$	0.9 kg/m^3	19	Uniform (0.6 to 1.2)
C_0	1.15 kg/m^3	50	Log-normal
D_0	w/c = 0.4 to 14.2 mm^2/year	75	Log-normal
	w/c = 0.5 to 41.0 mm^2/year		
	w/c = 0.6 to 86.4 mm^2/year		
	w/c = 0.7 to 162.7 mm^2/year		

x	1.0 cm	50	Normal
	2.0 cm		
	3.0 cm		
	4.0 cm		
	5.0 cm		
	6.0 cm		

Nogueira *et al.*

The evolution of probability of corrosion initiation in function of cover depth and w/c ratio, considering C.A. II, is presented in Figures 7, 8, and 9.

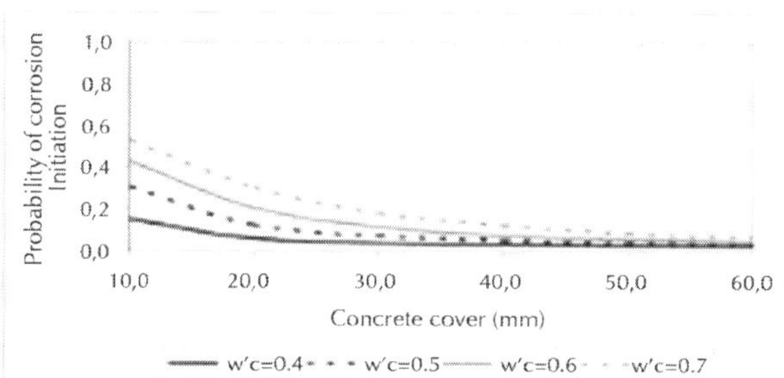

Figure 7: Evolution for probability of corrosion initiation. C.A. II structural lifetime equal to 5 years.

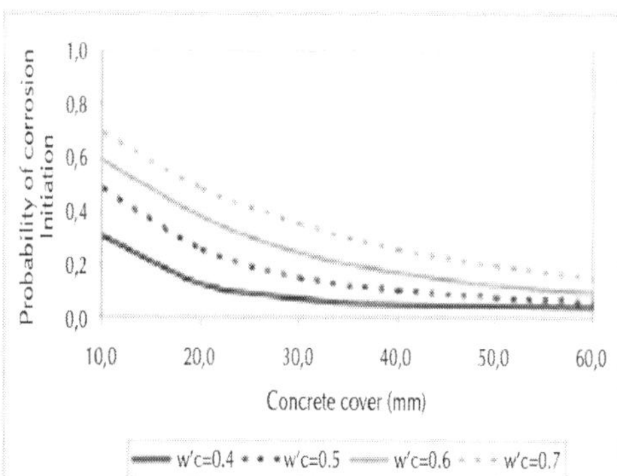

Figure 8: Evolution for probability of corrosion initiation. C.A. II structural lifetime equal to 15 years.

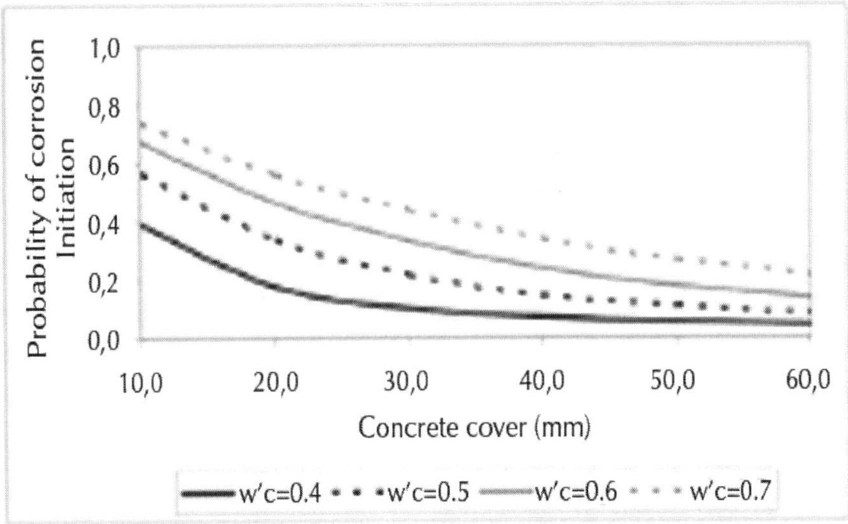

Figure 9: Evolution for probability of corrosion initiation. C.A. II structural lifetime equal to 25 years.

As presented on the three figures above, the probability of corrosion initiation reduces faster when the cover thickness increases from 10 to 40 mm. However, this reduction tends to stabilize when the cover depth is larger. Then, these results show that by increasing the cover thickness, an effective improvement on the structural safety is not observed from a certain value of depth. From this value, the structural cost grows faster than the structural safety against the corrosion initiation failure. However, this type of behavior is different for each value of w/c ratio adopted and the time considered. For concretes with low value of w/c, 0.4 to 0.5 for instance, the stabilization process is observed faster than in the permeable concretes, with w/c ratio of 0.6 or 0.7. Considering permeable concretes, thicker covers are required in order to achieve the stabilization on the reliability index value. The analyses aiming the determination of the evolution of probability of corrosion initiation in the function of cover depth and w/c ratio, involving C.A. III, are presented in Figures 10, 11, and 12.

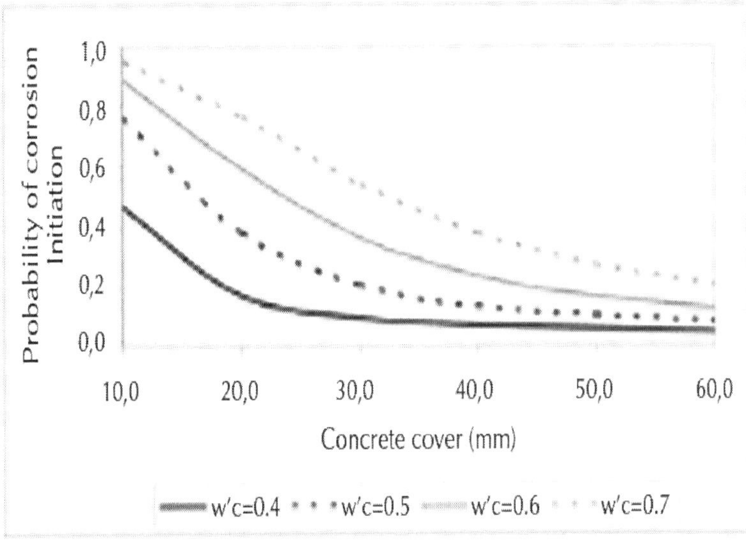

Figure 10: Evolution for probability of corrosion initiation. C.A. III structural lifetime equal to 5 years.

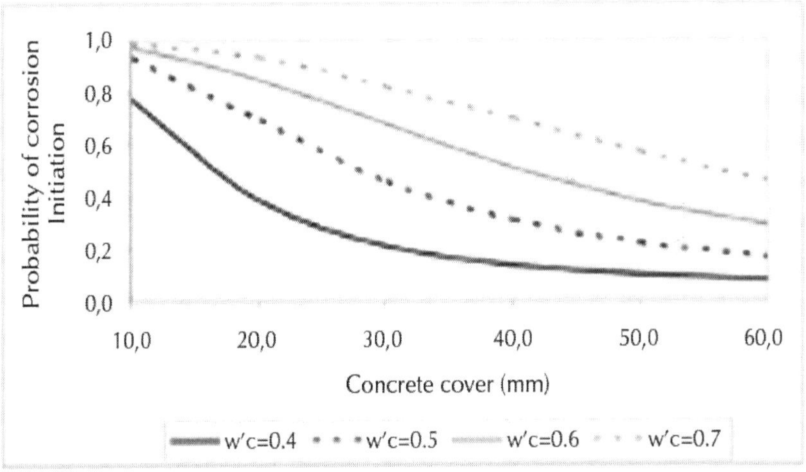

Figure 11: Evolution for probability of corrosion initiation. C.A. III structural lifetime equal to 15 years.

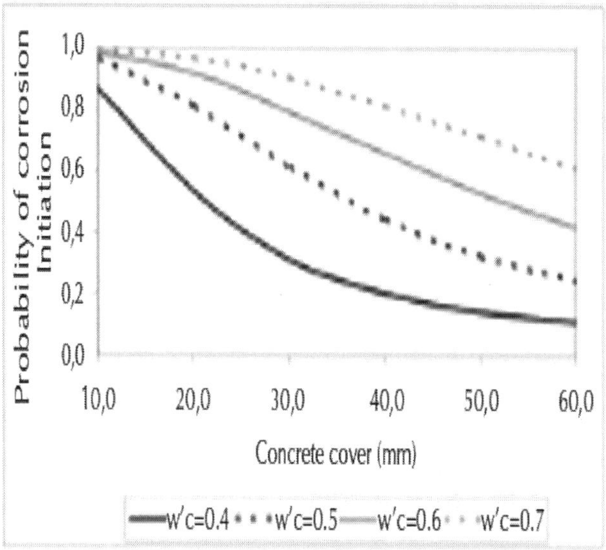

Figure 12: Evolution for probability of corrosion initiation. C.A. III structural lifetime equal to 25 years.

The behavior of curves observed in the above three figures is similar to those presented for C.A. II. However, due to the high environment aggressiveness, the probability of corrosion initiation tends to be higher than in C.A. II for the same structural lifetime and cover depth. The stabilization of the probability of corrosion initiation values, observed on previous analyses, also occurs for C.A. III. However, this process, for C.A. III, is observed for cover depth values thicker than in C.A. II case.

Another important aspect that should be mentioned is the time dependency considered on time corrosion initiation analyses. As long is the time desired to avoid the corrosion initiation or even though to reduce the probability of corrosion initiation process, the covers adopted must be thicker to reach the condition of reliability index stabilization. Thus, this type of analysis gives an interesting parameter which can be used to establish periods of structural maintenance according the type of concrete used and the environment where the structure is located.

The results presented in this application demonstrate the importance of stating correctly a value for the cover depth. As seen in this example, the cover depth is essential for structural durability, especially in reinforced concrete structures.

Example 3

The proposed probabilistic model can be used to determine the intervals of time in order to plan inspection and maintenance procedures in reinforced concrete structures as function of w/c ratio adopted. For each w/c value adopted for concrete design, an abacus can be constructed using the probabilistic model. These abacuses allow the choice of an optimum value of concrete cover in order to perform maintenance procedures in fixed time intervals. It is important to stress that these intervals are based on safety criterion. Then, the costs involved in maintenance and repair costs are not considered.

For this proposal, a reference value (named target) of reliability index was chosen, which was assumed as minimum requirement of safety. Then, when the structure has a reliability index lower than the target one, maintenance is required. According to JCSS (2001), the reliability index target expresses the requirements of structural safety, which guarantees acceptable values of structural failure risk considering a given scenario. In this regard, JCSS (2001) defines a set of values for reliability index target considering failure and serviceability limit state scenarios. The corrosion time initiation is considered as a serviceability limit state scenario. In this study, the loss of reinforcement cross-sections have not been considered, which is a failure limit state. Therefore, the reliability index target adopted is 1.3, leading a probability of failure approximately equal to 10%. Other values of target reliability index can, evidently, be adopted as function of the structural risk desired.

The analyses in this application were carried out considering the random variables presented in Table 2, as well as its statistical properties. Two classes of aggressiveness, C.A. II and C.A. III, were considered. The abacuses involving reliability index, time, and cover depth for C.A. II are presented in Figures 13 and 14. It is worth to mention that only w/c ratios of 0.4 and 0.5 were analyzed since higher w/c ratio led to reliability indexes lower than the target. Consequently, it indicates that these w/c ratios should not be used in reinforced concrete applications.

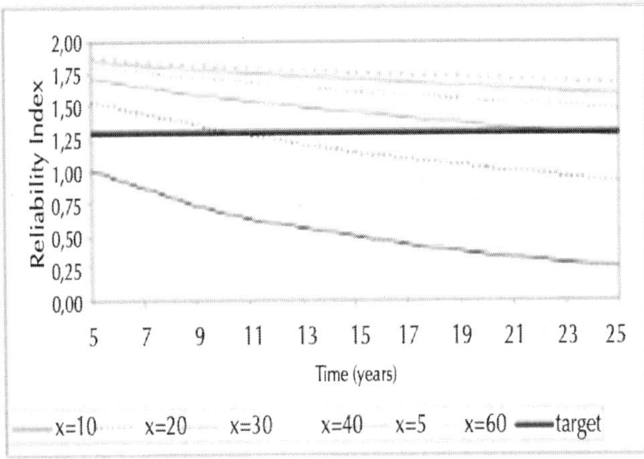

Figure 13: Abacus reliability index required for C.A. II and w/c ratio of 0.4.

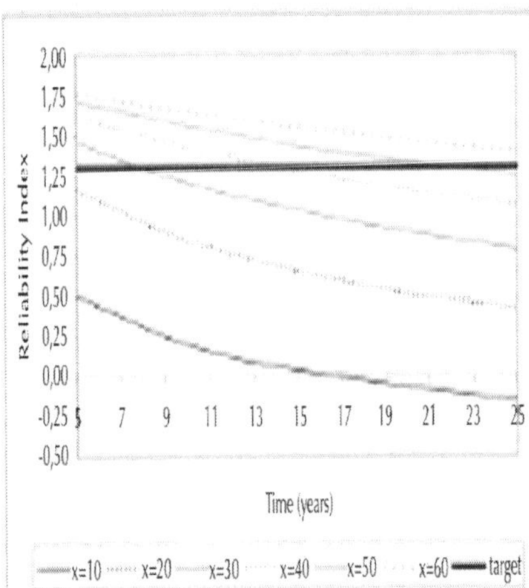

Figure 14: Abacus reliability index required for C.A. II and w/c ratio of 0.5.

Figure 13 presents the behavior of the reliability index in function of time for concretes with w/c ratio equal to 0.4 and C.A. II. According to these results, cover depth value of 10 mm is not recommended because it indicates failure (reliability index lower than the target) even for short time after construction. For cover depth of 20 mm, the maintenance procedures should be performed 10 years after construction. At this time, the structural reliability index is equal to the target. After this time, the structural reliability becomes lower than the target, indicating failure. It is worth to mention that cover depth values higher than 30 mm lead maintenance time interval of higher than 25 years. In these cases, analysts should study the possibility of change of w/c ratio in order to allow, at least, one inspection in the first 25 years of structural lifetime.

Similar behavior can be observed in Figure 14, where concretes with w/c ratio of 0.5 and C.A. II were considered. In this case, the minimum cover required is 30 mm which lead to maintenance procedures of 8 years after construction. Otherwise, considering cover depth equal to 50 mm, the maintenance procedure should be performed 21 years after construction in order to keep the structural safety at a desired level. Therefore, by changing the cover depth from 30 to 50 mm, or increasing the cover in 20 mm, the maintenance can be performed 13 years later.

On the other hand, considering cover depth of 30 mm, the maintenance time for concretes with w/c ratios of 0.4 and 0.5 are 23 and 8 years, respectively, for C.A. II. Then, an increment of 187.5% on the maintenance time interval can be verified by slightly changing the w/c ratio. Moreover, keeping the w/c ratio (0.5 for instance) constant and changing the cover depth from 30 to 40 mm, the maintenance times are 8 and 14 years, respectively, for C.A. II. Therefore, by changing the cover thickness by 10 mm, it results an improvement of 75% on the maintenance time. The abacuses involving the reliability index in function of w/c ratio and cover depth values for category of aggressiveness III are presented in Figures 15 and 16.

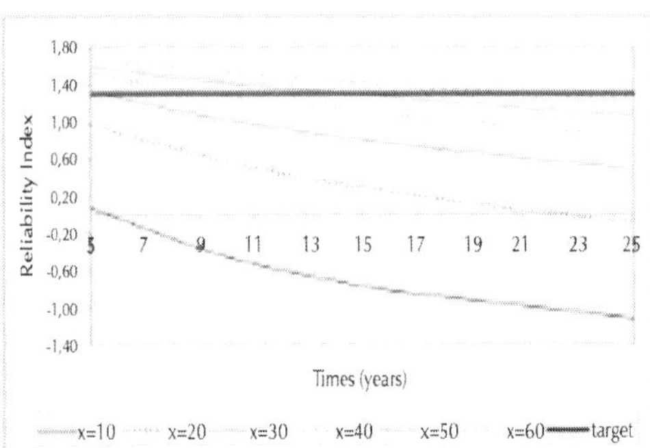

Figure 15: Abacus reliability index required for C.A. III and w/c ratio of 0.4.

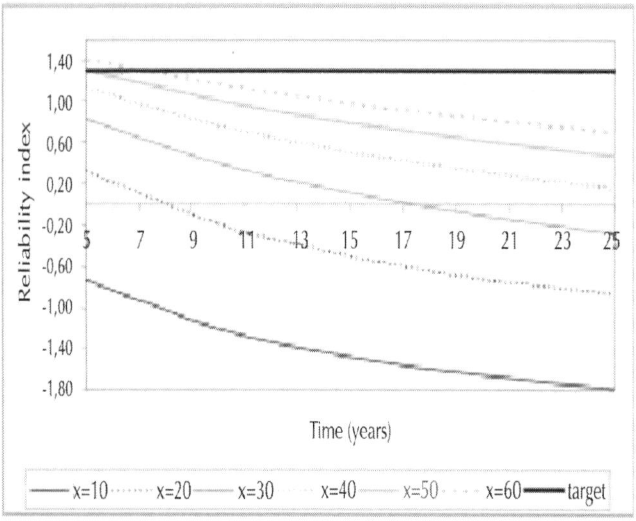

Figure 16: Abacus reliability index required for C.A. III and w/c ratio of 0.5.

Considering the two figures above, the same remarks discussed for C.A. II can be performed. However, in this case, the intervals of time for maintenance are considerable reduced due to the aggressiveness of the environment analyzed, which is higher than that observed for C.A. II.

It is important to emphasize that when the w/c ratio increases, small values of cover depth tend to be prohibitive regarding the corrosion time initiation failure. Moreover, in this case, the time interval for maintenance procedures will be drastically reduced. Therefore, regarding concretes with high values of w/c ratio, maintenance procedures in short periods of time and high values of cover depth have to be used in order to prevent reinforcement's corrosion.

Based on the results presented in this application, optimum concrete mixtures and cover depth values can be obtained in order to guarantee the durability of reinforced concrete structures. Moreover, these parameters can be allied aiming the evaluation of construction, maintenance, and reparation costs. Therefore, the global structural cost, taking into account the corrosion failure scenario, can be accurately achieved.

Example 4

As last part of this study, sensitivity analysis involving all random variables present on the probabilistic problem was performed. This example aims, therefore, at verifying which variables have more influence on the probability of corrosion initiation for each failure scenario considered. This example considers the random variables, as well as its statistical properties (presented on Table 2 and used in examples 2 and 3. Figures 17, 18, 19, and 20 show the influence of each random variable on the probability of corrosion initiation for C.A II, w/c ratios of 0.4 and 0.7, and structural lifetimes 5 and 25 years. These values of w/c ratio and structural lifetime were chosen because they are on the boundary of all analyses previously performed. The intermediate values of w/c and structural lifetime were not presented since they describe the transition behavior between the results obtained and shown below.

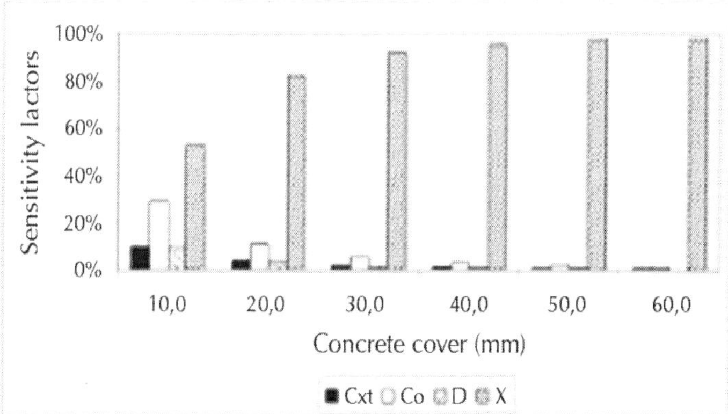

Figure 17: Sensitivity factors for C.A. II, w/c = 0.4 at 5 years of structural lifetime.

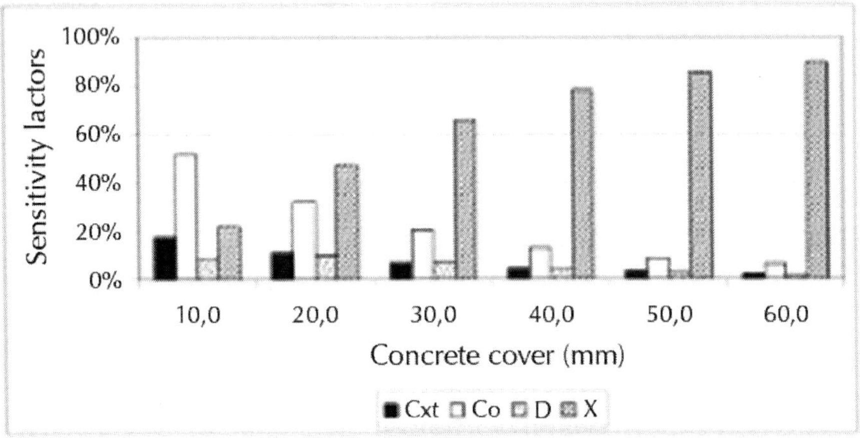

Figure 18: Sensitivity factors for C.A. II, w/c = 0.4 at 25 years of structural lifetime.

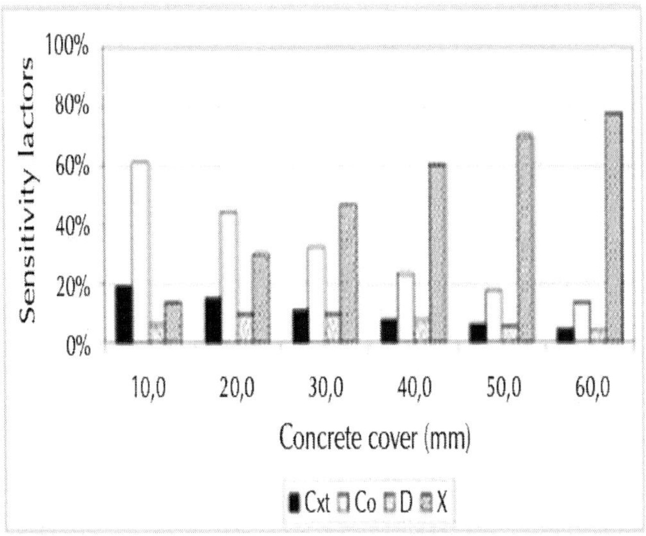

Figure 19: Sensitivity factors for C.A. II, w/c = 0.7 at 5 years of structural lifetime.

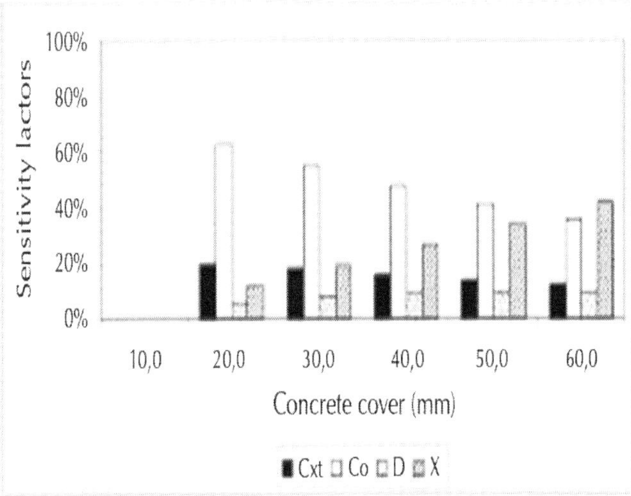

Figure 20: Sensitivity factors for C.A. II, w/c = 0.7 at 25 years of structural life time.

Regarding the C.A. II, it can be observed that for concretes with low w/c ratio, the more influent parameter is the cover depth, especially when thick covers are analyzed. The chloride concentration at the structural surface is another influent parameter. Its influence appears when thin covers are considered. Moreover, as the w/c ratio grows, the influence of C_0 tends to become higher even for thick covers. This behavior was expected since the concrete resistance against chloride ingress reduces when w/c ratio grows. Then, the structural resistance depends, in most part, on the concrete cover, which is required in order to equilibrate the effects of C_0.

The sensibility results for all random variables analyzed, considering C.A. III, are presented in Figures 21, 22, 23, and 24. These figures show that the cover depth has major importance when the environment aggressiveness increases. This influence is observed for all range of values considered for cover thickness.

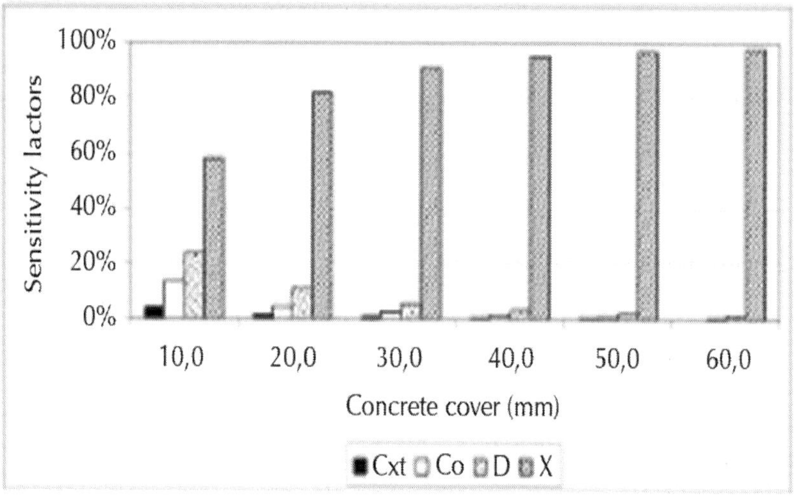

Figure 21: Sensitivity factors for C.A. III, w/c = 0.4 at 5 years of structural lifetime.

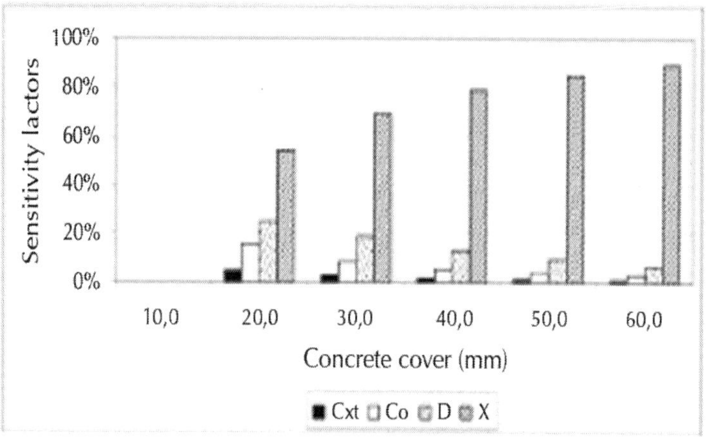

Figure 22: Sensitivity factors for C.A. III, w/c = 0.4 at 25 years of structural lifetime.

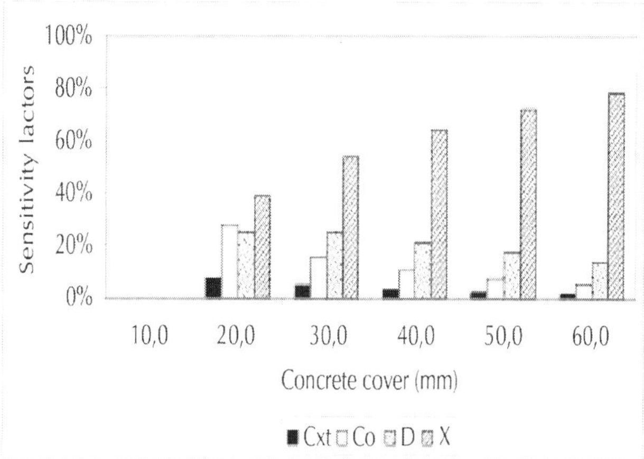

Figure 23: Sensitivity factors for C.A. III, w/c = 0.7 at 5 years of structural lifetime.

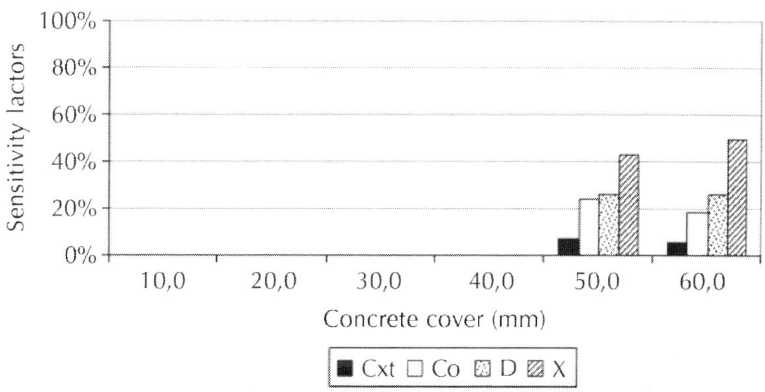

Figure 24: Sensitivity factors for C.A. III, w/c = 0.7 at 25 years of structural lifetime.

Based on the results presented in this example, the improvement on constructive techniques applied to reduce the randomness on the cover depth is justified. When the uncertainties present on this variable are reduced, the probability of corrosion initiation can be considerable improved. Then, the durability assessment can be performed accurately.

It is worth to mention that the blank spaces (no columns) presented in these four last figures indicate that the probabilistic analyses were carried out using Monte Carlo simulation. Then, in these cases, the sensibility quantification is not possible.

CONCLUSIONS

This paper presented a study of probabilistic corrosion initiation in reinforced concrete structures using the coupling among Fick's diffusion law and reliability algorithms. In this case, the corrosion starts when the chemical protection surrounding the steel bars is lost due to chloride ingress. The analyses involving the determination of probability of corrosion initiation along time were carried out considering two different categories of environment aggressiveness and several combinations of cover depth and concrete properties (w/c ratio).

As verified on the results presented, the corrosion initiation depends on several parameters. However, the chloride concentrations at the structural surface and the cover depth values have demonstrated to have major importance on probabilistic analysis. Equally, the coefficient of diffusion is also important in order to measure the material resistance against the chloride ingress, and it can be determinant for choosing the optimal values for concrete cover.

According the results achieved in this paper, the durability of reinforced concrete structures has to be associated with more fair values of cover depth, which are, in general, based only on international concrete standard design. This parameter is more realistically evaluated using probabilistic approaches, in order to take into account the inherent randomness present on degradation phenomenon, which affects structural durability.

Regarding the choice of optimum values of cover depth, it was observed that this parameter strongly depends on the concrete quality against porosity, which is reflected by w/c ratio, as well as on the period of time attributed for structural interventions for maintenance procedures. Evidently, the determination of optimum combination among concrete material mixture, cover depth value, and construction/maintenance/reparation costs is the next question to be answered in order for the structural design to become more economic, safe, and rational.

Based on the abacuses achieved using the proposed probabilistic model, which are based only on safety criterion, it can be observed that the use of strong concretes (i.e., with low w/c ratio) may become more interesting alternatives against simply cheaper concretes with high w/c ratio, where thick covers have to be adopted in order to prevent corrosion initiation. Therefore, based on the results presented, advanced analyses can be performed considering the conception, maintenance, and reparation costs in order to take into account the engineering solution which allies the w/c ratio and the cover depth value that lead an adequate level of structural safety.

AUTHORS' CONTRIBUTIONS

EDL developed the mechanical model based on Fick's law. Based on this model, CGN coupled reliability algorithms and performed all probabilistic analyses. EDL and HBC were the supervisors of this study and the organizers of the article. All authors read and approved the final manuscript.

AUTHORS' INFORMATION

CGN is a civil engineer, and nowadays, he works and in the post-doctorate level at the University of São Paulo. EDL is an assistant professor at the same university and a member of the editorial board of Engineering Analysis with Boundary Elements. HBC is a full professor at the same university and the head of GMEC research group.

ACKNOWLEDGMENTS

Sponsorship of this research project by the São Paulo State Foundation for Research (FAPESP) is greatly appreciated. This research is a part of the activities scheduled by the research project USP/COFECUB 2012.1.672.1.0.

REFERENCES

1. Angst U, Elsener B, Larsen CK, Vennesland Ø: Critical chloride content in reinforced concrete—a review. Cem Concr Res 2009, 39:1122-1138.
2. Balaji Rao K, Anoop MB, Lakshmanan N, Gopalakrishnan S, Appa Rao TVSR: Risk-based remaining life assessment of corrosion affected reinforced concrete structural members.,J Struct Eng 2004, 31:51-64.
3. Brazilian Association of Technical Standards (ABNT): NBR 6118: design of concrete structures—procedures. ABNT, Rio de Janeiro; 2003.
4. Crank J: The mathematics of diffusion. 2nd edition. Clarendon, Oxford; 1975.
5. Dhir RK, Jones MR, Ng SLD: Prediction of total chloride content profile and concentration/time-dependent diffusion coefficients for concrete. Mag Concr Res 1998, 50(37–48):1998.
6. Ditlevsen O, Madsen HO: Structural reliability method. Wiley, New York; 1996.
7. DuraCrete: Statistical quantification of the variables in the limit state functions. Technical report. The European Union-Brite EuRam III; 2000. Contract BRPR-CT95-0132, project BE95-1347, Report Number BE95-1347/R7, May, 2000
8. Enright MP, Frangopol DM: Probabilistic analysis of resistance degradation o reinforced concrete bridge beams under corrosion. Eng Struct 1998, 20:960-971.
9. Estes AC, Frangopol DM: Updating bridge reliability based on bridge management systems visual inspection results. J Bridge Eng 2003, 8:374-382.
10. Guzmán S, Gálvez JC, Sancho JM: Cover cracking of reinforced concrete due to rebar corrosion induced by chloride penetration. Cem Concr Res 2011, 41:893-902.
11. Hasofer AM, Lind NC: Exact and invariant second moment code format.
12. J Eng Mech Div 1974, 100(1):111-121.

13. Joint Committee on Structural Safety (JCSS): Probabilistic model code. JCSS, Denmark; 2001.
14. Jung WY, Yoon YS, Sohn YM: Predicting the remaining service life of land concrete by steel corrosion. Cem Concr Res 2003, 33:663-677.
15. Kirkpatrick TJ, Weyers RE, Anderson-Cook CM, Sprinkle M: Probabilistic model for the chloride-induced corrosion service life of bridge decks. Cem Concr Res 2002, 32:1943-1960.
16. Kong JS, Ababneh AN, Frangopol DM, Xi YP: Reliability analysis of chloride penetration in saturated concrete. Probabilist Eng Mech 2002, 17:305-315.
17. Leonel ED, Venturini WS: Multiple random crack propagation using a boundary element formulation. Eng Fract Mech 2011, 78:1077-1090.
18. Leonel ED, Chateauneuf A, Venturini WS, Bressolette P: Coupled reliability and boundary element model for probabilistic fatigue life assessment in mixed mode crack propagation. Int J Fatigue 2010, 32:1823-1834.
19. Leonel ED, Beck AT, Venturini WS: On the performance of response surface and direct coupling approaches in solution of random crack propagation problems.
20. Struct Saf 2011, 33:261-274.
21. Leonel ED, Venturini WS, Chateauneuf A: A BEM model applied to failure analysis of multi-fractured structures. Eng Fail Anal 2011, 18:1538-1549.
22. Leonel ED, Chateauneuf A, Venturini WS: Probabilistic crack growth analyses using a boundary element model: applications in linear elastic fracture and fatigue problems. Eng Anal Bound Elem 2012, 36:944-959.
23. Nogueira CG, Leonel ED, Coda HB: Reliability algorithms applied to reinforced concrete structures durability assessment. Ibracon Structures and Materials Journal 2012, 5(4):440-450.
24. Nowak AS, Collins KR: Reliability of structures. McGraw Hill, Michigan; 2000.
25. Papadakis VG, Roumeliotis AP, Fardis MN, Vagenas CG: Mathematical modeling of chloride effect on concrete durability and protection measures. In Concrete repair, rehabilitation and

protection. Edited by Dhir RK, Jones MR. E&FN Spon, London; 1996.

26. Rackwitz R, Fiessler B: Structural reliability under combined random load sequences. Comput Struct 1978, 9:489-494.
27. Sarveswaran V, Roberts MB: Reliability analysis of deteriorating structures—the experience and needs of practicing engineers. Struct Saf 1999, 21:357-372.
28. Stewart MG, Rosowsky DV: Structural safety and serviceability of concrete bridges subject to corrosion. J Infrastruct Syst 1998, 4:146-155.
29. Suo Q, Stewart MG: Corrosion cracking prediction updating of deteriorating RC structures using inspection information. Reliab Eng Syst Saf 2009, 94:1340-1348.
30. Thirumalai P, Ravi R, Parthiban GT: Potential monitoring system for corrosion of steel in concrete. Adv Eng Softw 2006, 37:375-381.
31. Tuutti K: Corrosion of steel in concrete. Swedish Cement and Concrete Research Institute, Stockholm; 1982.
32. Vu KAT, Stewart MG: Structural reliability of concrete bridges including improved chloride-induced corrosion models. Struct Saf 2000, 22:313-333.
33. Xiang T, Zhao R: Reliability evaluation of chloride diffusion in fatigue damaged concrete. Eng Struct 2007, 29:1539-1547.

Chapter 8

A Survey of High Performance Concrete Developments in Civil Engineering Field

Vatsal Patel[1], and Niraj Shah[2]

[1]G. H. Patel College of Engineering &Technology, Gujarat Technological University, Ahmedabad, India

[2]C. S. Patel Institute of Engineering & Technology, Charotar University of Science and Technology, Changa, India

ABSTRACT

High Performance concrete (HPC) has received increased attention in the development of infrastructure Viz., Buildings, Industrial Structures, Hydraulic Structures, Bridges and Highways etc. leading to utilization of large quantity of concrete. This paper presents a comprehensive coverage of High Performance concrete developments in civil engineering field. It highlights the High Performance concrete features and requirements over conventional concrete. Furthermore, recent trends with regard to High Performance Concrete development in this area are explored. This paper also includes effect of Mineral and Chemical Admixtures used to improve performance of concrete.

INTRODUCTION

Concrete is the most widely used construction material in India with annual consumption exceeding 100 million cubic meters. Conventional Ordinary Portland Cement Concrete which is designed on the basis of compressive strength does not meet many functional requirements as it is found deficit in aggressive environments, time of construction, energy absorption capacity, repair and retrofitting jobs etc. So, there is a need to design High Performance Concrete which is far superior to conventional concrete as the Ingredients of High Performance Concrete contribute most efficiently to the various properties. A High Performance Concrete is a concrete in which certain characteristics are developed for a particular application and environment so that it will give excellent performance in the structure in which it will be placed, in the environment to which it will be exposed, and with the loads to which it will be subjected during its design life [1].The strategic Highway Research Programmed (SHRP) has defined a High Performance Concrete as concrete meeting one of the following requirements:

1) 4 hours compressive strength ≥17.5 N/mm^2;
2) 24 hours compressive strength ≥35 N/mm^2;
3) 28 days compressive strength ≥70 N/mm^2;
4) Water cement ratio ≤0.35;
5) Durability factor >80 after 300 cycles of freezing and thawing.

High Performance Concrete is also defined as concrete which meets special performance and uniformity requirements that cannot be always achieved routinely by using conventional materials and normal mixing, placing and curing practices [2]. The requirements may involve enhancement of placement and compaction without segregation, long term mechanical properties, early age strength, volume stability or service life in severe environments. The ingredients generally used in investigations are 1) Cement; 2) Fine Aggregate; 3) Coarse Aggregate; 4) Water; 5) Mineral Admixtures; and 6) Chemical Admixtures [3-74].

ADMIXTURE APPLICATIONS IN HIGH PERFORMANCE CONCRETE

Admixtures play an important role in the production of High Performance Concrete. Mineral Admixtures form an essential part of the High Performance Concrete mix. They are used for various purposes depending upon their properties. Table 1 shows different types of mineral admixtures with their particle characteristics. Chemical composition determines the role of mineral admixtures in enhancing properties of concrete. Different materials with Pozzolanic properties such as Fly Ash (FA), Ground Granulated Blast Furnace Slag (GGBS), and Silica fume (SF), High Reactivity Met kaolin (HRM), Rice Husk Ash (RHA), Copper Slag, Fine Ground Ceramics have been

Table 1: Different mineral admixtures used in HPC

Mineral admixtures	Classification	Particle characteristics
Ground granulated blast furnace slag	Cementations and pozzolanic	Unprocessed materials are grain like sand, ground to size <45 μm particles and have a rough texture
Fly ash	Cementations and pozzolanic	Powder consists of particles size <45 μm, 10% to 15% are more than45 μm, solid spheres and generally smooth
Silica fume	Highly active pozzolana	Fine powder consisting of solid spheres of 0.1 μm average diameter
Rice husk ash	Highly active pozzolana	Particles are <45 μm in size and have cellular and porous structure

Widely used as supplementary cementations materials in the production of High Performance Concrete [3-74]. Such applications not only help to improve the strength and durability characteristics of High Performance Concrete but will also help to dispose more of the industrial by-products which are major environmental threats. Different Chemical admixtures (Super plasticizers) are extensively used in development of High Performance Concrete with very low water cement ratio are represented in Table 2 with their functions.

EXPERIMENTAL INVESTIGATIONS ON HIGH PERFORMANCE CONCRETE

Shannag et al. [2] prepared High Performance Concrete mixes with W/B ratio = 0.35 containing various proportions of natural pozzolan (blend of certain volcanic stuff) (0%, 5%, 10%, 15% by weight of cement) and silica fume (0, 20, 40, 60 kg/m^3) with naphthalene formaldehyde sulfonated superplasticizer and checked performance in severe sulphate environments (Na_2SO_4, $MgSO_4$) and in Sea Waters. The investigation indicated that the concrete mix containing 15% natural pozzolan, and 15% silica fume showed the best protection in sulphates solutions and sea waters by retaining more than 65% of its Compressive strength after one year of storage. High Performance concrete two way slabs cast with M60 grade concrete with W/C ratio 0.32% and 7.5% silica fume exhibit good strength than RCC slab [3]. High Performance reinforced concrete columns were manufactured at various replacement levels of silica fume (0%, 5%, 7.5%, and 10%) and flyash (10%) with superplasticizer CERAPLAST 300 and W/B ratio 0.3 to target a mean strength of M_{60}. The cube compressive strength results at 3 days, 7 days, 28 days, 56days and 90 days indicated that maximum strength was obtained with 5% SF and 10% FA at all ages [4]. The effect of solution on the compressive strength of HPC mixes was checked by Shakir A. Al-Mishhadani et al. [5]. Results of destructive and non-destructive tests were statistically

Table 2: Different chemical admixtures used in HPC

Chemical admixtures	Function
Superplasticizer	To reduce the water requirement by 15% to 20% without affecting the workability leading to a high strength and dense concrete
Accelerator	To reduce the setting time of concrete thus helping early removal of forms and therefore used in cold weather concreting
Retarder	To increase the setting time by slowing down the hydration of cement and therefore are preferred in places of high temperature concreting
Water reducing admixture	To achieve certain workability (slump) at low water cement ratio for a specified strength thus saving on the cement
Air entraining admixture	To entrain small air bubbles in concrete which act as rollers thus improving the workability and therefore very effective in freeze-thaw cycles as they provide a cushioning effect on the expanding water in the concreting in cold climate

Analyzed by using SPSS verson15 software and statistical models were proposed to evaluate compressive strength, splitting tensile strength, modulus of rupture and static modulus of elasticity. High Performance Concrete mixes containing different percentages of met kaolin were tested for strength and durability and have shown better resistances to the attacks of chemicals such as chloride and sulphates when exposed to these chemicals for 180 days period [6]. Durability in terms of Chloride ion permeability was measured for metakaolin based High Performance Concrete [7]. Bhanja et al. [8] developed silica fume based high Performance Concrete mixes and found significant improvements in the tensile strength. Durability factor at the end of 30 freezing and thawing cycle of High Performance Concrete samples were examined [9]. Rheological properties of High Performance Concrete mixes produced by rice husk ash and fly ash were studied and found that for low yield stress and moderate plastic viscosity, blending of equal masses of silica fume and rice husk ash seems to be a suitable admixture [10]. A. H. Menon et al. [11] studied effect of mineral admixtures like fly ash, ground granulated blast furnace slag, silica fume and chemical admixtures (super plasticizers) on the porosity, pore size distribution and compressive strength development of high strength concrete in seawater curing condition. The High Performance Concrete was prepared by adding tire rubber fibers with composition of concrete at 0.25 W/B ratio [12]. The compressive strength at 3 days, 7 days and 28 days was reduced by 20% for High Performance Concrete with rubber tyre in comparison with control High Performance Concrete. A High Performance Concrete of grade M30 was prepared by replacing cement with ground granulated blast furnace slag (0%, 40%, 50%, 60%), sand with robo sand (crusher dust) (0%, 5%, 10%, 15%, 20%, 25%, 30%), 1.5% SAVEMIX SP111 admixture and tested for compressive and tensile strengths at 7 days and 28 days [13]. The percentage increase of compressive strength of concrete is 11.06% and 17.6% at the age of 7 and 28 days by replacing 50% of cement with GGBS and 25% of sand with ROBO sand. Pazhani et al. [14] produced High Performance Concrete by replacing 20%, 40%, 60%, 80%, 100% of fine aggregate with copper slag and 30% of cement with GGBS and tested to assess the durability parameters such as water absorption and chloride ion penetration. The water absorption and chloride ion penetrability for 30% replacement of cement with GGBS were decreased by 4.58% and 29.90% respectively & for 100% replacement of fine aggregate with copper slag same were decreased by 33.59% and 77.32% respectively. The mechanical and durability properties of different High Performance Concrete mixes containing fly ash were studied and have been used to develop precast concrete products such as non-pressure pipes and heavy duty paver blocks and these developments are reported [15]. A. Camoes et al. [16] produced low cost High Performance concrete by replacing fly ash with cement. Workability, Mechanical and durability properties of the produced concrete were studied.

Hani H. Nassif et al. [17] evaluated the elastic modulus of high performance concrete made from mixes using various percentages of fly ash, silica fume and granulated blast furnace slag. The effect of curing methods namely air-dry curing, curing compound and wet curing with burlap on the elastic modulus was also studied. An experimental work was carried out regarding the basic physical characteristics, mechanical and fracture-mechanics properties, durability charecteristics, hydric and thermal properties of high performance concrete with upto 60% of Portland cement replaced by fine ground ceramics [18]. K. E. Hassan et al. [19] carried laboratory study on the properties of superplasticised high performance concrete by using SF and FA (10%, 30% by weight of cement). The SF concrete showed similar strength development to that of the Ordinary Portland Cement concrete but slight higher values at all tested ages (1, 3, 7, 28, 365 days). FA concrete gave lowest compressive strength at early ages, same at 28 days and higher at 365 days than OPC concrete. SF and FA reduced the permeability by 87% and 84% respectively in the long term (365 days). A study was carried out on the properties of Portland cement concrete (PCC), High volume fly ash concrete (HFAC, containing 40% of FA) and concrete (GGFAC, containing a combination of 25% FA and 15% GGBS) by assessing compressive strength and resistance to H_2SO^4 attack. The compressive strength of GGFAC increased by 23.3%, higher than that of PCC, but lower than that of HFAC between 28 days and 1 year.The compressive strength of GGFAC and HFAC increased with increasing immersion period in H_2SO_4 solution. The combination of FA and GGBS improved both short and long term properties of concrete, while HFAC requires a relatively longer time to get its beneficial effect [20]. Abhilash Shukla et al. [21] evaluated the optimum percentage (0%, 5%, 10%, 15%, and 20%) of Rice Husk Ash as a partial replacement of cement for M30 and M60 grade of concrete. There was a significant improvement in compressive strength (3% to 10% increase) and flexural strength (0.6% to 8% increase) of the concrete with rice husk content of 10% for M30 and M60 grade concrete at 7 days and 28 days. By using linear regression technique M. F. Zain et al. [22] suggested formulae that relate splitting tensile strength with that of compressive strength, W/B ratio and concrete age for High Performance Concrete. The STS values obtained from experiment compared well with that estimated from this formula, with the average ratio of the experimental/predicted data being close to unity and proposed to be used to estimate the STS of High Performance Concrete. The effect of coarse aggregate type (crushed quartzite, crushed granite, lime stone and marble) on the compressive strength, splitting tensile strength, fracture energy, characteristic length and elastic modulus of concrete produced with 28 days target compressive strengths of 30 (W/B = 0.55), 60 (W/B = 0.44) and 90 MPa (W/B = 0.26) was studied [23] and found that for all grades crushed quartzite coarse aggregates concrete showed higher compressive strength (44.8 MPa, 70.4 MPa, 98.2 MPa), higher split tensile strength (4.1 MPa, 5.2 MPa, 8.4 MPa), higher fracture energy (143.1 MPa, 158.1 MPa, 166.1 MPa), higher modulus of elasticity (37.5 GPa, 39.5 GPa, 48.2 GPa) and higher characteristic length (0.319 m, 0.231 m, 0.113 m). A High Performance Concrete has been successfully developed for a digitally-controlled printing process which can build architectural and structural components without formwork. The well printed concrete had significantly fewer voids, higher flexural strength [24]. B. K. Raghu Prasad et al. [25] proposed an artificial neural network (ANN) to predict 28 days compressive strength of high performance concrete. The high values of R2 demonstrated that the proposed ANN model was suitable for predicting the compressive strength values very closely with the experimental values. With extremely low W/B (0.14 - 0.18), high binder content, multi addition of SF, GGBS, lime stone powder and high standard superplasticizer Ultra High performance concrete can be prepared with common technology and without removing the coarse aggregate [26]. Compressive strength, Chloride Permiability and shrinkage cracking potential test results indicate that admixtures may enhance durability without increasing strength [27]. Grade 100 high performance Concrete (W/B = 0.223) was produced by using sulphate resisting cement and silica fume (10%) with 1% superplasticizer for the marine environment. It showed high impermeability (stable current of circuit 2.7 - 3.5 mA, destroyed time 41 days), high elastic modulus (6.5 MPa), high ability to steel adhesion (steel adhesion factor = 33.37), high strength development (60 MPa) at 3 days [28]. A. Elahi et al. [29] carried out investigation to evaluate the mechanical and durability properties of High Performance Concrete (W/B = 0.3) containing supplementary cementations materials (Silica Fume, Fly Ash, Ground Granulated Blast Furnace Slag) in binary and ternary systems. Portland cement was replaced with fly ash up to 40%, silica fume up to 15% and GGBS upto a level of 70%. The ternary mixes containing GGBS or Fly Ash (50%) and Silica Fume (7.5%) performed the best amongst all the mixes to resist the chloride diffusion. Silica fume (7.5%) performs better than other supplementary cementations materials for the strength development. There was a benefit in case of sorptivity with the addition of 20% FA and 50% GGBS compared to control concrete. High Performance Concrete mixes containing fine ground ceramics as Portland cement replacement in an amount of up to 60% of mass were developed by Eva Vejmelkova et al. [30]. The compressive strength decreased very fast for the replacement levels higher than 20%; the produced concrete lost its high performance character. For same level water absorption coefficient was relatively low, the frost resistance was excellent and showed good chemical resistance in $MgCl_2$, NH_4Cl, Na_2SO_4, HCl environments. H. S. wong et al. [31] carried laboratory investigation to determine the cementing efficiency factor k of a pozzolan (number of parts of cement in a concrete mixture that could be replaced by one part of pozzolan) without changing the property. Investigation on silica fume (SF) and metakaoline (MK) concrete found that k varied with pozzolan type, replacement level and age. K increased with age but declined with high pozzolanic content. H. Z. Lopez-Calvo et al. [32] evaluated the influence of calcium nitrite based corrosion inhibitor (CNI) at addition rates of 0, 12.5 and 25 L/m3 on the compressive strength of High Performance Concrete manufactured with 8% silica fume blended cement in combination with 0%, 20% and 40% FA replacements and mixed at 0.29, 0.37 and 0.45 W/B ratio. In concretes tested at 28 days and 1 year, when 12.5 L/m3 CNI was incorporated into the mixture, an increase in compressive strength of approximately 15% to 17% were observed for all the W/B ratio. In concrete specimens tested at 9 years, when 12.5 L/m3 CNI was added to a concrete mixture, an increase in compressive strength of 4%, 13%, 23% for 0.29, 0.37 and 0.45 W/B was found. An experimental investigation was carried out by Da Xu et al. [33] to develop High Performance Concrete with targeted strength 120 MPa. Two mixes of OPC concrete were cast, one with 10% Unidensified Silica fume (USF) replacement and other one with 10% USF and 30% Ultra-Fine Ground Granulated Blast Furnace Slag (UFGGBS) replacement with 0.25 W/B ratio. Both Mix showed High Compressive Strength at 28 days (as high as 135 MPa), a high flexural strength (nearly 12 MPa), Very high secant modulus of elasticity (45.5 MPa), Electrical resistivity above 20 $k\Omega\text{-}^{cm}$ result in a low corrosion rate for the embedded steel reinforcement. Khalifa S. Al Jabri et al. [34] investigated the effect of copper slag as a replacement of sand on the properties of high performance concrete (W/B = 0.35). Eight concrete mixtures were prepared with different proportions of copper slag (0%, 10%, 20%, 40%, 50%, 60%, 80%, and 100%). Beyond 50% of copper slag as sand replacement caused reduction in strength and increase in water absorption rate. Results showed that 40% weight of copper slag can be used as replacement of sand in order to obtain High Performance Concrete with good properties. High Performance Concrete with a maximum compressive strength of 95.4MPa was produced using 10% metakaolin with W/B ratio of 0.3 [35]. The Chloride Ion Permeability value decreased considerably with increase in metakaolin content from 0 to 30% thus indicating improved durability with increasing metakaolin content. Dr. H. Sudarsana Rao et al. [36] produced Glass fibre reinforced High Performance Concrete (W/B = 0.35) with Fly Ash (0%, 10%, 20%, 30%) as the mineral admixture and glass fibre (0%, 0.5%, 1%, 1.5%). Cubes and Cylinders were tested to assess compressive and tensile strength characteristics and also workability by conducting compaction factor test. Compaction factor workability of GFRHPC mixes was decreased with increase in percentage replacement of cement by fly ash for all percentages of glass fibre ratio. Maximum compressive strength and tensile strength was attained at 1% of glass fibre ratio with 10% fly ash. Sung Won Yoo et al. [37] evaluated autogenous shrinkage in HPC (W/B = 0.3) with mineral admixtures fly ash (0%, 10%, 15%, 20%, 30%) and silica fume (0%, 5%, 7.5%, 10%, 15%) & chemical admixture shrinkage reducing agent and expansion agent. The autogenous shrinkage (AGS) in High Performance Concrete with fly ash was decreased continuously with larger fly ash replacement. AGS in High Performance Concrete with silica fume was increased compared to that in OPC concrete. Use of both admixtures in adequate amount can provide decrease in AGS as well as improvement of the strength. J. Brozovsky et al. [38] indicated calibration correlations as elaborated for high performance concrete class C55/67 to C80/95; 2 to 60 days old, for Schmidt impact hammer type N&L. They are compared with relations specified in the norms EN13791 and CSN731373, and concluded that they are applicable to practical detection of built in high performance concrete strength. Noumowe et al. [39] carried out experimental investigations to study the effect of elevated temperature (600°C) on permeability of control high performance concrete, high performance concrete incorporating polypropylene fibres and high performance concrete made with lightweight aggregates. The W/B ratio was kept 0.3 and silica fume was used 10% by weight of cement as replacement. A heating-cooling cycle was applied on cylindrical specimens; 65 mm thick slices were cut from each cylinder and dried prior to being subjected to permeability test (before and after the heating cooling cycle). There was severe loss in the permeability of High Performance Concrete with normal aggregates. The permeability of the fibre reinforced high performance concrete was greater than that of the control high performance concrete. Investigations were carried out by K. M. A. Hossain et al.

[40] to assess the suitability of using volcanic ash (VA) as a cement replacement material to produce High Performance Concrete having constant W/B ratio of 0.3. HPVACs having a 28-day compressive strength in excess of 60 MPa can be achieved by using upto 20% VA. The drying shrinkage of HPVACs was slightly higher compared to control concrete (0% VA). All HPVACs showed rapid chloride permeability (RCP) values within the range of 100 - 1000 Coulomb which suggests that they have very low permeability as per ASTM C 1202 - 97. There was no obvious increase in the width of the Interfacial transition zone in HPVACs compared to control concrete. Three point bend tests for High Performance Concrete notched beams were performed to determine the fracture parameters by Ricardo A. Einsfeld et al. [41]. The results obtained by the work of fracture method apparently follow the trend of fracture energy increasing as the compressive strength of the concrete increases. The results for fracture energy obtained by the size effect method showed a slight trend to decrease with increasing compressive strength. Compressive strength, gas permeability and carbonation of High Performance Concrete (W/B ratio 0.25, 0.30 and 0.35) with fly ash (15%, 30%, 45%, and 60%) or ground granulated blast furnace slag (15%, 30%, 45%, 60%) were experimentally investigated by Shi Hui Sheng et al. [42]. At lower W/B ratio, compared to control High Performance Concrete, FA with replacement upto 30% improved compressive strength, and had marginal increase in both gas permeability and carbonation depth. High Performance Concrete with GGBS showed better performance and properties were little affected by W/B ratio. The compressive strength of silica fume (10%, 20%, and 30%) concretes was investigated by E. H. Kadri et al. [43] at lower water binder (0.25, 0.30, 0.35, and 0.45) with a naphthalene sulphonate superplasticizer. The results showed that partial cement replacement upto 20% had produced higher compressive strengths than control concretes for all W/B ratios. An experimental investigation on the flexural behaviour of reinforced high performance concrete was conducted by Paramsivam suresh kumar et al. [47] by using crushed sand stone as fine and coarse aggregates in addition with silica fume and fly ash combined with superplasticizer.

The beams were made with concrete having compressive strength in the range of 74 - 88 N/mm^2 and tensile reinforcement in the range of 1.34% to 3.14%. The ultimate experimental moment was found to be 3% - 15% higher than predicted ultimate moment by ACI 318, actual deflections of the beams were found slightly above allowable values under service loads & observed crack widths under service loads were within acceptable limits. Effect of kerosene and Gasoline on some properties of high performance concrete (W/B = 0.31) which can be used as storage tanks for petroleum products was checked by performing experimental work. Specimen exposed to Kerosene and Gasoline showed more weight loss and reduction in compressive strength at 7, 28 and 90 days [48]. By using fractional factorial design concept, Salilkumar roy et al. [49] prepared sixteen mixes from a basic mix (cement: sand:aggregates:fly ash = 1:1.3:2.6:0.8 with W/C = 0.37) and determined changes (cement = 0.1; sand = 0.1; aggregates = 0.2; fly ash = 0.4 and changing water cement ratio by 0.01) using Taguchi's orthogonal array. The changed composition mix was found to have highest compressive strength and lowest water absorption and therefore found to be best for making paving blocks. Flexural and Spliting tensile strengths of high performance fiber reinforced concrete were determined with variables fiber volume fraction (0%, 0.5%, 1%, 1.5% with an aspect of 80), silica fume replacement level (0%, 10%, 20%, 30%) and W/B ratio = 0.35. The addition of steel fibers upto 1.5% resulted in increase of 38% in the flexural tensile strength, and an increase of 56% in the splitting tensile strength compared with a plain concrete matrix [50]. K. Arunachalam et al. [51] performed experiments to make a comparative study on the properties of high performance concrete (W/B = 0.28) with fly ash (25% and 50% replacement) and without fly ash (control concrete) in normal and aggressive environment (Al_2SO_4 and NaCl). In case of control concrete under aggressive environment compressive strength was decreased from 28 to 60 days by about 25% of 28 days strength. There was a continuous increase in compressive strength from 7 days to 60 days for high performance fly ash concrete (both 25% and 50% cement replacement) for both environment. Five categories of High Performance Concrete namely Normal Strength Concrete (35 MPa), Medium Strength Concrete (56 MPa), High Strength Concrete (62 MPa), Very High Strength Concrete (98 MPa), Ultra High Strength Concrete (120 MPa) with slump value ranged from (100 - 200 mm) were designed by Mohamed M Yoursy E l shikh [52] with the help of commercial software taking into account the Egyptian practice and assigned that to different column cases (edge, corner and internal) for high rise residential building (12 stories with 3m height of each storey).

The columns are analysed and designed which showed that total cost reduction of all columns vary from (32% to 57%) corresponding to concrete strength (56 to 100MPa) compared with Normal Strength Concrete (35 MPa). Vaishali G Ghorpade [53] performed tests on four mixes of concrete with 0%, 0.5%, 1.0% and 1.5% by volume fraction of glass fiber, silica fume (0%, 10%, 20%, 30% by weight of cement) with W/B ratio = 0.35, aggregate/binder = 2.0 and superplasticizer 1% of the weight of cement. The optimum percentage recommended as 1% fibre volume with 10% silica fume for achieving maximum benefits in compressive strength, split tensile strength and flexural strength. Experimental investigations were carried out by Saeed ahmed al. Sheikh [54] on High Performance Concrete (W/B = 0.35) by heating them in an electric furnace to varying temperatures upto 1000°C and the change in compressive strength, tensile strength, ultrasonic pulse velocity and rebound hammer were determined. High Performance Concrete showed reduction in compressive and tensile strengths about 69% and 73% respectively. Pulse Velocity and rebound hammer of High Performance Concrete were also decreased by 76% and 21% respectively. Quantitative assessment of different cement replacement levels with silica fume (0%, 2.5%, 5%, 7.5%, 10%, 12.5%, 15%) on strength and durability properties for M60 (W/B = 0.32), M70 (W/B = 0.3) and M110 (W/B = 0.232) grades of High Performance Concrete mixes was carried out by K. Perumal et al. [55]. Replacement level of 10% with silica fume in M_{60}, M_{70}, and M_{110} grades of High Performance Concrete mixes was found to be the optimum level to obtain higher values of compressive strength (72.22 MPa, 82.46 MPa, 121.22 MPa), split tensile strength (6.10 MPa, 5.85 MPa, 8.45 MPa), flexural strength (8.3 MPa, 9.1 MPa, 8.45 MPa) and elastic modulus (35.1 GPa, 37.5 GPa, 44.4 GPa) & lower values of porosity (2.2%, 2.15%, 1.95%), sorptivity (0.0210 $mm/min^{0.5}$, 0.0205 $mm/min^{0.5}$, 0.0104 $mm/min^{0.5}$) and saturated water absorption (1.25%, 1.2%, 1.05%). Compressive Strength and Modulus of Elasticity of High Performance Concrete made with silica fume, ground granulated blast furnace slag, fly ash and normal Portland cement with W/B = 0.25 and exposed to temperature within the range of 20°C to 50°C under different types of curing methods were determined [56]. The highest levels of strength and modulus of elasticity were obtained by silica fume concrete under water and wrapped curing at temperature of 35°C. High Performance Concrete was produced by Andres Salas et al. [57] by using Rice

Husk Ash samples (conventional type TRHA obtained by a thermal treatment and other chRHA was based on a chemical thermal attack to the rice husk ash). Compressive and Flexural Strengths of chRHA concrete were comparable to a silica fume concrete made with the same replacement level & these strengths were higher than the control and TRHA mixtures.

Through the optimization of the proportions of compositions and employing heat treatment, very high performance concrete with large quantities of ultrafine mineral powders such as pulverized fly ash, pulverized granulated blast furnace slag and silica fume was successfully prepared which showed compressive strength upto 200 MPa [58]. High Performance concrete utilizing fly ash and microsilica as cement replacing materials was investigated [59] to study permeation related properties like permeability, porosity and sorptivity with different curing conditions namely moist curing, air curing and curing at 5°C. The permeability, porosity and sorptivity values were lowest for low W/B ratio with 10% microsilica and fly ash replacement in the range of 15% - 20%. Same permeability values were not achieved for specimens cured at 5°C in comparison with moist and air curing.

The research was conducted [60] to obtain the optimum High Performance Concrete (W/B = 0.35) mix by using 10% silica fume and various percentages of pulvarized fly ash (40%, 60%, 80%) with melamine superplasticizer. The resistance of the produced concrete when exposed to an acidic environment (H_2SO_4, HCl) was reduced in mix with 10% silica fume and fly ash more than 60% as an OPC replacement. Compressive strength of High Performance Concrete and Paste mixes was found by K. O. kjellsen et al. [61] at ages 1 day to 4 years with W/B ratio ranged from 0.25 to 0.4 and upto 10% of the cement was replaced by silica fume. The concretes and the pastes with 10% silica fume appeared to reduce strength between 3 and 9 months and were higher at 2 years than after 3 months. The direct tensile strength, flexural strength, compressive strength, dynamic modulus of elasticity and fracture energy of High Performance Concrete mixes (W/B = 0.25 & 0.4) and silica fume (0%, 10%) at 28 days and 2 years were studied by K. O. Kjellsen et al. [62]. An increase of 20% to 25% in direct tensile strength and 10% to 25% in flexural & compressive strength with little effect on the dynamic modulus of elasticity was found with 10% silica fume.

A method to find coefficient of thermal expansion of High Performance Concrete was proposed by H. Kada et al. [63] in which concrete samples instrumented with vibrating wire extensometer were given thermal shocks & measured deformation and temperature signal were used to calculate coefficient of thermal expansion. High Performance Concrete with 0%, 5%, 10% and 20% metakaolin at W/B ratio of 0.3 was prepared by C. S. Poon et al. [64] to investigate compressive strength, porosity and pore size distribution and compared with concrete made by 5% and 10% silica fume as replacement of cement & cured in water at 27°C for 3 days to 90 days. Metakaolin replaced concrete showed higher compressive strength at all days compared to silica fumed concrete. Porosity and pore diameter was decreased for all concrete as curing age increased. A High Performance C40 grade non dispersible underwater concrete was manufactured with the use of Non dispersible admixture by Jiang Cong-Sheng et al. [65] which showed low flow loss, low strength loss and excellent anti dispersion in comparison with control conctere. PAN Zhihua et al. [66] developed high performance foamed concrete using blast furnace slag, fly ash and silica fume as cement replacement by means of pre foaming process with different mix proportions of mineral admixtures which showed compressive strength 1.1 - 23.7 MPa and thermal conductivity of about 0.16 - 0.75 W/m°C. It was observed that with small amount of superplasticizer the compressive strength reached as high as 44.1 MPa. Lu Jianfu et al. [67] examined compressive strength & chloride penetration resistance of high performance concrete under three curing conditions namely SC (standard curing), MC (Matched curing, sealed specimens with plastic membrane in temperature controlled curing box) and MNC (Matched curing of naked concrete, naked specimens in temperature controlled curing box). The concrete under MC and MNC showed better resistance to chloride penetration than one under SC and compressive strength of SC concrete was higher at 248 days. Corrosion Reinforced high performance concrete was studied by Pedro Montes et al. [68] by using fly ash and calcium nitrite corrosion inhibitor. The combination of good quality concrete (W/C = 0.29) with Calcium Nitrite Corrosion Inhibitor at an addition rate of 12.5 L/m^3 was found best to reduce the effect of Chloride induced corrosion of steel reinforcement. Hardened properties of the high performance concrete containing high replacement level of slag (50% - 80%) under air and wet curing regimes were studied by Erhan

Guneyisi et al. [69]. With increase in slag content, corrosion resistance was increased, compressive strength, splitting tensile strength and water absorption were decreased. Rice husk Ash with high carbon High Performance Concrete [70] showed superior performance in the mechanical and durability tests compared with the reference mix. The mixing time of the High Performance Concrete [71] can be reduced by increasing the relative solid concentration (by optimising the particle size distribution, replacement of cement by silica fume), matching the type of superplasticizer and cement in the mix and increasing the speed of mixing. A HPC made with different W/C ratios (0.30, 0.40) and light weight sand (0% and 30% by volume) as sand substitution by Soo Duck Hwang [72] was used to check the effect of moist curing. Combined use of 30% light weight sand and 7 days of moist curing can lead to greater increase in compressive strength and larger decrease in total shrinkage compared to the use of 30% light weight sand without moist curing or 7 days of moist curing without light weight sand. K.chinnaraju et al. [74] conducted an Experimental test programme to study the effect of fly ash, GGBS and silica fume on High Performance Concrete with W/B ratio 0.32. The control concrete mixture of grade M60 included only ordinary Portland cement as binder and other mixtures incorporated binary (PC + flyash, PC + GGBS, C + SF), ternary (PC + flyash + GGBS, C + flyash + SF, PC + GGBS + SF) and quaternary (PC + flyash + GGBS + SF) cementitious blends were designed. The replacement levels for both flyash and GGBS were 10%, 20% and 30%, while those of SF were 2.5%, 7.5%, 10% and 12.5% by weight of cement as addition. The mix with 20% replacement of cement by 10% of fly ash and 10% of silica fume showed the best compressive strength for 7 days and 28 days, with an increase in compressive strength of 15.64% and 13.24% respectively. The increase in splitting tensile strength was moderate for various combinations of fly ash and GGBS with silica fume upto 10%. Beyond that there was no increase in splitting tensile strength. The change in flexural strength was limited for most of the combinations but was decreased with higher rate beyond the addition of 10% silica fume. Several researchers designed High Performance Concrete mixes by using mix design methods of Normal Strength Concrete or by combining principles of one or more methods. However, suitability of existing mix design methods such as Indian Standard Code, Department of Environment, American Concrete Institute etc. for designing High Performance Concrete mixes giving necessary modifications had not been reported so far. From the experimental trials it was observed that all the methods of mix design did not give desired workability and compressive strength and needed some modifications for making High Performance Concrete mixes. Kumbhar et al. [44] suggested following modifications for existing mix design methods to achieve High Performance Concretes.

- Indian Standard code method gives lesser proportion of fine aggregates so a proper course to fine aggregate ratio should be recommended for better workability of mix. Further, Indian Standard code method incorporates use of only one mineral admixture. There should be modification in method in terms of addition of more than one mineral admixture with appropriate proportion of chemical admixture.
- Department of Environment method suggests higher quantity of fine aggregate which increases water required and hence reduces the strength which is not acceptable for High Performance Concrete. Therefore proper dosage of Super plasticizer should be recommended.
- Modified American Concrete Institute method gives usage of High dosage of Super plasticizer to get high slump which reduces the strength therefore proper content of SP should be recommended.

A mix design procedure for High Performance Concrete had been suggested by aminul Islam Lascar [45] which took rheological parameters into account to determine compressive strength, water cement ratio and aggregate volume to paste volume ratio. Instead of using water cement ratio and compressive strength relationship, relationship between compressive strength, paste volume-aggregate volume ratio; physical properties of aggregate and rheological parameters were used in mix design. Correlation charts for rheological parameters and compressive strength was developed based on cube test results of several trial mixes whose rheological parameters had also been found by the remoter. Muhammad Fauzi Mohd.Zain et al. [46] developed prototype expert system called HPCMIX according to the mix design method proposed by Aitcin that gave proportion of trial mix of High Performance Concrete and recommendations on mix adjustment. Mehta and Aitcin [73] proposed a simplified mixture proportioning procedure that was applicable for normal weight concrete with compressive strength values between 60 and 120 MPa.

This method was suitable for coarse aggregates having a maximum size of between 10 mm and 15 mm and slump values of between 200 and 250 mm.The optimum volume of aggregate was suggested to be 65% of the volume of the high performance concrete.

CONCLUSIONS

1) High Performance Concrete can be prepared to give optimized performance characteristics for a given loading and exposure conditions along with the requirements of cost, service life and durability.
2) The applications of concrete will necessitate the use of High Performance Concrete incorporating new generation chemical admixtures (PCE based superplasticizers) and available mineral admixtures.
3) The success of High Performance Concrete requires more attention on proper Mix Design, Production, Placing and Curing of Concrete. For each of these operations controlling parameters should be achieved by concrete producer for an environment that a structure has to face.

REFERENCES

1. Kmita, "A New Generation of Concrete in Civil Engineering," Journal of Materials Processing Technology, Vol. 106, No. 1-3, 2000, pp.80-86. doi:10.1016/S0924-0136(00)00642-7
2. M. J. Shannag and H. A. Shaia, "Sulfate Resistance of High-Performance Concrete", Cement & Concrete Composites, Vol. 25, No. 3, 2003, pp. 363-369. doi:10.1016/S0958-9465(02)00049-5
3. N. V. Ramana, K. Gnaneswar, C. Sashidhar and T. Naresh Kumar, "Behavior of High Performance Concrete Two Way Slabs in Punching Shear", International Journal of Science and Advanced Technology, Vol. 2, No. 3, 2012, pp. 122-126.
4. P. Muthupriya, K. Subramanian and B. G. Vishnuram, "Experimental Investigation on High Performance Reinforced Concrete Column with Silica Fume and Fly Ash as Admixtures",

Asian Journal of Civil Engineering (Building and Housing), Vol. 12, No. 5, 2011, pp. 597-618.

5. S. A. Al-Mishhadani, W. A. Al-Qaisi and S. F. Al-Khafaji, "Predicting Mechanical Properties of High Performance Concrete by Using Non-Destructive Tests," Journal of Engineering & Technology, Vol. 27, No. 3, 2009, pp. 425-444.
6. B. Patil and P. D. Kumbhar, "Strength and Durability Properties of High Performance Concrete Incorporating High Reactivity Metakaolin," International Journal of Modern Engineering Research, Vol. 2, No. 3, 2012, pp. 1099-1104.
7. V. G. Ghorpade and H. S. Rao, "Chloride Ion Permeability Studies of Metakaolin Based High Performance Concrete," International Journal of Engineering Science and Technology ,Vol. 3, No. 2, 2011, pp. 1617-1623.
8. S. Bhanja and B. Sengupta, "Influence of Silica Fume on the Tensile Strength of Concrete," Cement and Concrete Research, Vol. 35, No. 4, 2005, pp. 743-747. doi:10.1016/j.cemconres.2004.05.024
9. K. Yildiz and L. O. Ugur, "Examinatıon of Durablity of High Performance Concrete (hpc) That Has Been Subjected to MgSO4 and NaCl Corrosıon against Freezing and Thawing," Scientific Research and Essay, Vol. 4, No. 9, 2009, pp. 929-935.
10. A I. Laskar and S. Talukdar, "Rheological Behavior of High Performance Concrete with Mineral Admixtures and Their Blending," Construction and Building Materials, Vol. 22, No. 12, 2008, pp. 2345-2354. doi:10.1016/j.conbuildmat.2007.10.004
11. H. Memon, S. S. Radin, M. F. M. Zain and J. F. Trottier, "Effects of Mineral and Chemical Admixtures on High-Strength Concrete in Seawater," Cement and Concrete Research, Vol. 32, No. 3, 2002, pp. 373-377. doi:10.1016/S0008-8846(01)00687-1
12. R. F. Martins and J. L. Akasaki, "Study of Compressive Strength of High Performance Concrete Added with Tire Rubber", Materials Compuestos, 2005, pp. 1-8.
13. V. Malagavelli and P. N. Rao, "High Performance Concrete with Ggbs and Robo Sand," International Journal of Engineering Science and Technology, Vol. 2, No. 10, 2010, pp. 5107-5113.

14. Pazhani and R. Jeyaraj, "Study on Durability of High Performance Concrete with Industrial Wastes," Applied Technologies & Innovations, Vol. 2, No. 2, 2011, pp. 19- 28.
15. A. Peter, M. Neelamegam, J. K. Dattatreya, N. P. Rajamane and S. Gopalkrishnan, "Utilisation of Flyash as Cement Replacement Material to Produce High Performance Concrete", In: B. Cltatterjee, K. K. Singh & N. G. Goswami, Eds., Fly Ash Utilisation for Value Added Products, 1999, pp. 38-49.
16. A Camoes, B. Aguiar and S. Jalali, "Durability of Low Cost High Performance Fly Ash Concrete," International Ash Utilization Symposium, Center for Applied Energy Research, University of Kentucky, Paper 43, 2003.
17. H. H. Nassif , H. Najm and N. Suksawang, "Effect of Pozzolanic Materials and Curing Methods on the Elastic Modulus of HPC," Cement & Concrete Composites, Vol. 27, No. 6, 2005, pp. 661-670. doi:10.1016/j.cemconcomp.2004.12.005
18. E. Vejmelkova, M. Keppert , P. Rovnanikova, M. Ondracek, Z. Kersner and R. Cerny, "Properties of High Performance Concrete Containing Fine-Ground Ceramics as Supplementary Cementitious Material," Cement & Concrete Composites, Vol. 34, No. 1, 2012, pp. 55-61. doi:10.1016/j.cemconcomp.2011.09.018
19. E. Hassan, J. G. Cabrera and R. S. Maliehe, "The Effect of Mineral Admixtures on the Properties of High Performance Concrete," Cement & Concrete Composites, Vol. 22, No. 4, 2000, pp. 267-271. doi:10.1016/S0958-9465(00)00031-7
20. G. Li and X. Zhao, "Properties of Concrete Incorporating Fly Ash and Ground Granulated Blast-Furnace Slag," Cement & Concrete Composites, Vol. 25, No. 3, 2003, pp. 293-299. doi:10.1016/S0958-9465(02)00058-6
21. A Shukla, C. K. Singh and A. Sharma, "Study of the Properties of Concrete by Partial Replacement of Ordinary Portland Cement by Rice Husk Ash," International Journal of Earth Sciences and Engineering, Vol. 4, No. 6, 2011, pp. 965-968.
22. F. M. Zain, H. B. Mahmud, A. Ilham and M. Faizal, "Prediction of Splitting Tensile Strength of High-Performance Concrete," Cement and Concrete Research, Vol. 32, No. 8, 2002, pp. 1251-1258. doi:10.1016/S0008-8846(02)00768-8

23. K. Wu, B. Chen, W. Yao and D. Zhang, "Effect of Coarse Aggregate Type on Mechanical Properties of High-Performance Concrete," Cement and Concrete Research, Vol. 31, No. 10, 2001, pp. 1421-1425. doi:10.1016/S0008-8846(01)00588-9
24. T. T. Le, S. A. Austin, S. Lim, R. A. Buswell, R. Law, A. G. F. Gibb and T. Thorpe, "Hardened Properties of HighPerformance Printing Concrete," Cement and Concrete Research, Vol. 42, No. 3, 2012, pp. 558-566. doi:10.1016/j.cemconres.2011.12.003
25. K. Raghu Prasad, H. Eskandari and B. V. Venkatarama Reddy, "Prediction of Compressive Strength of SCC and HPC with High Volume Fly Ash Using ANN," Construction and Building Materials, Vol. 23, No. 1, 2009, pp. 117-128. doi:10.1016/j.conbuildmat.2008.01.014
26. Wang, C. Yang, F. Liu, C. Wan and X. Pu, "Preparation of Ultra-High Performance Concrete with Common Technology and Materials," Cement & Concrete Composites, Vol. 34, No. 4, 2012, pp. 538-544. doi:10.1016/j.cemconcomp.2011.11.005
27. S. P. Shah and W. J. Weiss, "High Performance Concrete: Strength, Permeability and Shrinkage Cracking," International Symposium on High Performance Concrete, 2000, pp. 331-340.
28. P. H. Hanh and N. V. Tuan, "High Performance Concrete Used for Marine Gravity Concrete Works," The 3rd ACF International Conference, 2008, pp. 598-606.
29. A Elahi, P. A. M. Basheer, S. V. Nanukuttan and Q. U. Z. Khan, "Mechanical and Durability Properties of High Performance Concretes Containing Supplementary Cementitious Materials," Construction and Building Materials, Vol. 24, No. 3, 2010, pp. 292-299. doi:10.1016/j.conbuildmat.2009.08.045
30. Vejmelkova, M. Keppert, P. Rovnanikova, M. Ondracek, Z. Kersner and R. Cerny, "Properties of High Performance Concrete Containing Fine Ground Ceramics as Supplementary Cementitious Material," Cement & Concrete Composites, Vol. 34, No. 1, 2012, pp. 55-61. doi:10.1016/j.cemconcomp.2011.09.018
31. H. S. Wong and H. A. Razak, "Efficiency of Calcined Kaolin and Silica Fume as Cement Replacement Material for Strength Performance," Cement and Concrete Research, Vol. 35, No. 4, 2005, pp. 696-702. doi:10.1016/j.cemconres.2004.05.051

32. H. Z. Lopez-Calvo, P. Montes-Garcia, T. W. Bremner, M. D. A. Thomas and V. G. Jiménez-Quero, "Compressive Strength of HPC Containing CNI and Fly Ash after Long-Term Exposure to a Marine Environment," Cement & Concrete Composites, Vol. 34, No. 1, 2012, pp. 110- 118. doi:10.1016/j.cemconcomp.2011.08.007

33. Xu, B. S. Divsholi, D. T. Y. Lim and S. Teng, "Mechanical Properties and Durability of High Performance Concrete Incorporating Ultra Fine Slag and Undensified Silica Fume," 36th Conference on Our World in Concrete & Structures Singapore, 14-16 August 2011, pp. 1-10.

34. K. S. Al-Jabri, M. Hisada, S. K. Al-Oraimi and A. H. Al-Saidy, "Copper Slag as Sand Replacement for High Performance Concrete," Cement & Concrete Composites, Vol. 31, No. 7, 2009, pp. 483-488. doi:10.1016/j.cemconcomp.2009.04.007

35. V. G. Ghorpade and H. S. Rao, "Chloride Ion Permeability Studies of Metakaolin Based High Performance Concrete," International Journal of Engineering Science and Technology, Vol. 3, No. 2, 2011, pp. 1617-1623.

36. H. S. Rao, H. M. Somasekharaiah and V. G. Ghorpade, "Strength and Workability Characteristics of Fly Ash Based Glass Fibre Reinforced High Performance Concrete," International Journal of Engineering Science and Technology, Vol. 3, No. 8, 2011, pp. 6266-6277.

37. S. W. Yoo, S. J. Kwon and S. H. Jung, "Analysis Technique for Autogenous Shrinkage in High Performance Concrete with Mineral and Chemical Admixtures," Construction and Building Materials, Vol. 34, 2012, pp. 1-10. doi:10.1016/j.conbuildmat.2012.02.005

38. J. Brozovsky, D. Prochazka and D. Benes, "Determination of High Performance Concrete Strength by Means of Impact Hammer," The 10th International Conference of the Slovenian Society for Non-Destructive Testing, Ljubljana, 1-3 September, 2009, pp. 233-241.

39. A N. Noumowe, R. Siddique and G. Debicki, "Permeability of High-Performance Concrete Subjected to Elevated Temperature (600°C)," Construction and Building Materials, Vol. 23, No. 5, 2009, pp. 1855-1861. doi:10.1016/j.conbuildmat.2008.09.023

40. K. M. A. Hossain and M. Lachemi, "Strength, Durability and Micro-Structural Aspects of High Performance Volcanic Ash

Concrete," Cement and Concrete Research, Vol. 37, No. 5, 2007, pp. 759-766. doi:10.1016/j.cemconres.2007.02.014

41. R. A. Einsfeld and M. S. L. Velasco, "Fracture Parameters for High-Performance Concrete," Cement and Concrete Research, Vol. 36, No. 3, 2006, pp. 576-583. doi:10.1016/j.cemconres.2005.09.004
42. H.-S. Shi, B.-W. Xu and X.-C. Zhou, "Influence of Mineral Admixtures on Compressive Strength, Gas Permeability and Carbonation of High Performance Concrete," Construction and Building Materials, Vol. 23, No. 5, 2009, pp. 1980-1985. doi:10.1016/j.conbuildmat.2008.08.021
43. H. Kadri, S. Aggoun, S. Kenai and A. Kaci, "The Compressive Strength of High-Performance Concrete and Ultrahigh-Performance," Advances in Materials Science and Engineering, Vol. 2012, 2012, Article ID: 361857. doi:10.1155/2012/361857
44. P. D. Kumbhar and P. B. Murnal, "Assessment of Suitability of Existing Mix Design Methods of Normal Concrete for Designing High Performance Concrete Mixes," International Journal of Civil and Structural Engineering , Vol. 3, No. 1, 2012, pp. 158-167.
45. A Laskar, "Mix Design of High-Performance Concrete," Materials Research, Vol. 14, No. 4, 2011, pp. 429-433. doi:10.1590/S1516-14392011005000088
46. F. M. Zain, M. N. Islam and I. H. Basri, "An Expert System for Mix Design of High Performance Concrete," Advances in Engineering Software, Vol. 36, No. 5, 2005, pp. 325-337. doi:10.1016/j.advengsoft.2004.10.008
47. S. Kumar, M. A. Mannan and K. V. John, "High Performance Reinforced Concrete Beams made with Sandstone Reactive Aggregates," The Open Civil Engineering Journal, Vol. 36, No. 5, 2008, pp. 41-50.
48. N. Mahde, A. Jalawi and D. S. Atwan, "Effect of Kerosene and Gasoline on Some Properties of High Performance Concrete," Journal of Engineering, Vol. 17, No. 6, 2011, pp. 1643-1657.
49. S. K. Roy, H. Sugiharto, A. Kristanto and S. Himawan, "Systematic Formulation of High Performance Concrete Pavement," Civil Engineering Dimension, Vol. 7, No. 2, 2005, pp. 57-60.
50. Ramadoss and K. Nagamani, "Tensile Strength and Durability Characteristics of High Performance Fiber Reinforced Concrete,"

The Arabian Journal for Science and Engineering, Vol. 33, No. 2, 2008, pp. 307-319.

51. K. Arunachalam and R. Gopalakrishnan, "Experimental Investigation on High Performance Fly Ash Concrete in Normal and Aggressive Environment," 29th Conference on Our World in Concrete & Structures, Singapore City, 25-26 August 2004, pp. 181-188.
52. M. MY. El Shikh, "Economy of Using High Performance Concrete in Columns," 33rd Conference on Our World in Concrete & Structures, Singapore, 25-27 August 2008, pp. 1-8.
53. V. G. Ghorpade, "An Experimental Investigation on Glass Fibre reinforced High Performance Concrete with Silicafume as Admixture," 35th Conference on Our World in Concrete & Structures, Singapore, 25-27 August 2010, pp. 1-8.
54. S. A. Al. Sheikh, "Mechanical Properties for High Performance Concrete Exposed to High Temperature," International Journal of Civil and Structural Engineering, Vol. 2, No. 2, 2011, pp. 534-444.
55. K. Perumal and R. Sundararajan, "Effect of Partial Replacement of Cement with Silica Fume on the Strength and Durability Characteristics of High Performance Concrete," 29th Conference on Our World in Concrete & Structures, Singapore, 25-26 August 2004, pp. 397-404.
56. M. F. Mohd Zain and S. S. Radin, "Physical Properties of High-Performance Concrete with Admixtures Exposed to a Medium Temperature Range 20°C to 50°C," Cement and Concrete Research, Vol. 30, No. 8, 2000, pp. 1283- 1287. doi:10.1016/S0008-8846(00)00294-5
57. A Salas, S. Delvasto, R. Mejia de Gutierrez and D. Lange, "Comparison of Two Processes for Treating Rice Husk Ash for Use in High Performance Concrete," Cement and Concrete Research, Vol. 39, No. 9, 2009, pp. 773-778. doi:10.1016/j.cemconres.2009.05.006
58. Long, X. Wang and Y. Xie, "Very-High-Performance Concrete with Ultrafine Powders," Cement and Concrete Research, Vol. 32, No. 4, 2002, pp. 601-605. doi:10.1016/S0008-8846(01)00732-3

59. M. I. Khan, "Permeation of High Performance Concrete," Journal of Materials in Civil Engineering, Vol. 15, No. 1, 2003, pp. 84-92. doi:10.1061/(ASCE)0899-1561(2003)15:1(84)

60. A. K. Tamini, "High Performance Concrete Mix for an Optimum Protection in Acidic Conditions," Materials and Structures, Vol. 30, No. 3, 1997, pp. 188-191. doi:10.1007/BF02486392

61. K. O. Kjellsen, O. H. Wallevik and M. Hallgren, "On the Compressive Strength Development of High Performance Concrete and Paste Effect of Silica Fume," Materials and Structures, Vol. 32, No. 1, 1999, pp. 63-69. doi:10.1007/BF02480414

62. K. O. Kjellsen, M. Hallgren and O. H. Wallevik, "Fracture Mechanical Properties of High Performance Concrete—Influence of Silica Fume," Materials and Structures, Vol. 33, No. 9, 2000, pp. 552-558. doi:10.1007/BF02480535

63. H. Kada, M. Lachemi, N. Petrov, O. Bonneau and P.-C. Aitcin, "Determination of the Coefficient of Thermal Expansion of High Performance Concrete from Initial Setting," Materials and Structures, Vol. 35, No. 1, 2002, pp. 35-41. doi:10.1007/BF02482088

64. C. S. Poon, S. C. Kou and L. Lam, "Pore Size Distribution of High Performance Metakaolin Concrete," Journal of Wuhan University of Technology-Mater. Sci. Ed., Vol. 17, No. 1, 2002, pp. 42-46.

65. C.-S. Jiang, L.-N. Lu, S.-B. Guan, Q.-J. Ding and S.-G. Hu, "Preparation of High Performance Non-Dispersible Concrete," Journal of Wuhan University of TechnologyMater. Sci. Ed., Vol. 19, No. 2, 2004, pp. 67-69

66. Z. H. Pan, F. Hiromi and T. H. Wee, "Preparation of High Performance Foamed Concrete from Cement, Sand and Mineral Admixtures," Journal of Wuhan University of Technology-Mater. Sci. Ed., Vol. 22, No. 2, 2007, pp. 295-298. doi:10.1007/s11595-005-2295-4

67. J. F. Lü, H. Guan, W. X. Zhao and H. J. Ba, "Compressive Strength and Permeability of High-Performance Concrete," Journal of Wuhan University of Technology-Mater. Sci. Ed., Vol. 26, No. 1, 2011, pp. 137-141.

68. P. Montes, W. Theodore and B. F. Castellanos, "Interactive Effects of Fly Ash and CNI on Corrosion of Reinforced High-Performance

Concrete," Materials and Structures, Vol. 39, No. 2, 2006, pp. 201-210. doi:10.1617/s11527-005-9026-3

69. E. Guneyisi and M. Gesoglu, "A Study on Durability Properties of High-Performance Concretes Incorporating High Replacement Levels of Slag," Materials and Structures, Vol. 41, No. 3, 2008, pp. 479-493. doi:10.1617/s11527-007-9260-y
70. G. C. Cordeiro, R. D. T. Filho and E. M. Rego Fairbairn, "Use of Ultrafine Rice Husk Ash with High-Carbon Content as Pozzolan in High Performance Concrete," Materials and Structures, Vol. 42, No. 7, 2009, pp. 983-992. doi:10.1617/s11527-008-9437-z
71. O. Mazanec, D. Lowke and P. Schie, "Mixing of High Performance Concrete: Effect of Concrete Composition and Mixing Intensity on Mixing Time," Materials and Structures, Vol. 43, No. 7, 2010, pp. 357-365. doi:10.1617/s11527-009-9494-y
72. S.-D. Hwang, K. H. Khayat and D. Youssef, "Effect of Moist Curing and Use of Lightweight Sand on Characteristics of High-Performance Concrete," Materials and Structures, Vol. 46, No. 1-2, 2013, pp. 35-46. doi:10.1617/s11527-012-9881-7
73. P. C. Aitcin, "High-Performance Concrete," Taylor & Francis e-Library, 2004.
74. K. Chinnaraju, K. Subramanian and S. R. R. Senthilkumar, "Strength Properties of HPC Using Binary, Ternary and Quaternary Cementitious Blends," Structural Concrete, Vol. 11, No. 4, 2010, pp. 191-198. doi:10.1680/stco.2010.11.4.191

Chapter 9

Development of Optimal Structural System for Hybrid Cable-Stayed Bridges Using Ultra High Performance Concrete

Hee Seok Kim, Young Jin Kim*, Won Jong Chin, Hyejin Yoon

Structural Engineering Research Division, Korea Institute of Construction Technology, Goyang, Republic of Korea

ABSTRACT

This study developed an optimal structural system for the hybrid cable-stayed bridge expected to have a durable lifetime of 200 years and of which major structural members are made of ultra high performance concrete (UHPC) with 200 MPa-class compressive strength. This innovative cable-stayed bridge system makes it possible to reduce each of the construction and maintenance costs by 20% compared to the conventional concrete cable-stayed bridge by improving significantly the weight and durability of the bridge. Therefore, detail design is carried out considering a real 800 m cable-stayed bridge and the optimal structure of the hybrid cable-stayed bridge is proposed and verified.

INTRODUCTION

Recently, the construction market of long-span bridges in Korea has been boosted owing to the change in the land development paradigm. However, the Korean technological level related to long-span bridges remains still immature resulting in a strong reliance on advanced countries' core technologies for the construction of longspan bridges. Therefore, need is to implement R&D dedicated to long-span bridges to prevent further outflow of the national wealth and reduce the national budget expenditure.

In the case of cable-stayed bridge as one of the representative types of long-span bridges, concrete cablestayed bridges are known to be economically efficient for main spans running between 200 m and 400 m but can also be economically competitive in various aspects for spans longer than 400 m according to the selective exploitation of the materials. The construction of an economically efficient long-span bridge is primordially relying on the reduction of the weight. To that goal, considerable solutions can be reducing the weight of concrete itself or achieving lightweight structure by the reduction of the size through the use of concrete with higher strength. To date, efforts have been ceaselessly exerted to develop lightweight concrete to reduce the weight of concrete itself and applications on structural members are being under consideration. Besides, for the solution using concrete with higher strength, the Korea Institute of Construction Technology has secured original technology related to ultra high performance concrete (UHPC) with 200 MPa-class compressive strength and durable lifetime of 200 years [1-3]. However, the application of the developed UHPC to hybrid cable-stayed bridge is subordinated to the overcoming of the generally recognized limitations of concrete cable-stayed bridges, which have been rarely applied in Korea, together with the establishment of practical technologies of UHPC. To complicate this situation, the absence of cable-stayed bridge erected using UHPC wherever in the world necessitates the development of sectional shapes and structural systems fitted to UHPC for the application of this new material to the complex cable-stayed bridge structure and the establishment of component technologies and integrated technology enabling to secure structural safety and economic efficiency.

As a cable-stayed bridge having its main structural members made of UHPC with compressive strength of 200 MPa and durable lifetime of 200 years, the hybrid cable-stayed bridge system to be developed in this study is an innovative cable-stayed bridge system providing reduction of each the construction and maintenance costs by 20% compared to the conventional concrete cablestayed bridge by improving significantly the weight and durability of the bridge. Therefore, detail design is carried out considering a real 800 m cable-stayed bridge to derive the optimal structural system of the hybrid cablestayed bridge. Thereafter, the so-derived optimal system is checked for each construction stage and its stability is verified so as to develop the technology for UHPC cablestayed bridge.

DERIVATION OF THE OPTIMAL STRUCTURAL SYSTEM FOR THE HYBRID CABLE-STAYED BRIDGE

This study developed the optimal structural system for the reduced-cost and long-life hybrid cable-stayed bridge with main span 800 m using UHPC in order to construct original and economically efficient long-span bridges relying on our domestic technology.

The structural parameters influencing the performance of a cable-stayed bridge are more diversified than those of a common girder bridge. The performance, aesthetics and economic efficiency of the entire structural system depends on the structural system determined by each of these parameters. Even if there are no unique structural parameters, the values of the optimal structural parameters tend to vary sensitively according to the materials of the stiffening girder affecting directly the ratio of the permanent loads to the live loads.

In order to derive the optimal structural system fitted to the UHPC hybrid cable-stayed bridge, parametric analysis is conducted for the major parameters including the cable spacing, the side span length ratio and height of the pylons. The optimal structural system is finally derived by assessing the structural and economic efficiencies.

Parametric Evaluation for the Cable Spacing

The spacing of the main cables has critical influence not only on the structural efficiency during construction and in the completed system but also on the appearance of the bridge. In this evaluation, the main cables are realistically arranged at spacing of 7.0 m, 10.5 m and 14.0 m (Figures 1 to 3). Structural analysis is performed for the whole system considering each spacing to compare the stresses developed in the girder with regard to the loads governing the design that are the permanent loads, the live loads and the replacement and breakage of cables and determine the appropriate spacing of the cables. Since this evaluation focuses on the effects of the external forces, the prestressing tendons inside the cross-section of the hybrid cable-stayed bridge are not included in the models as being corresponding to internal forces.

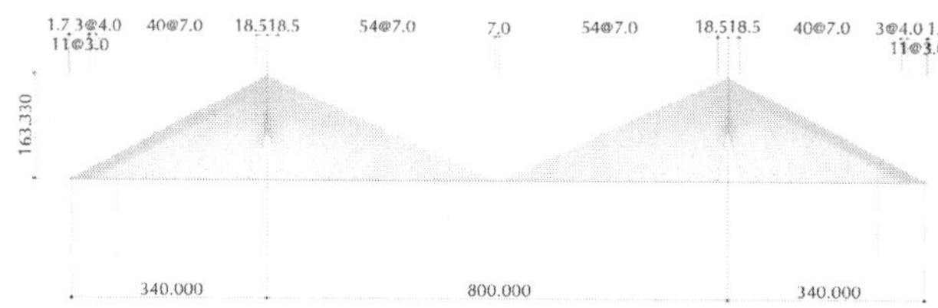

Figure 1: Cable spacing of 7.0 m.

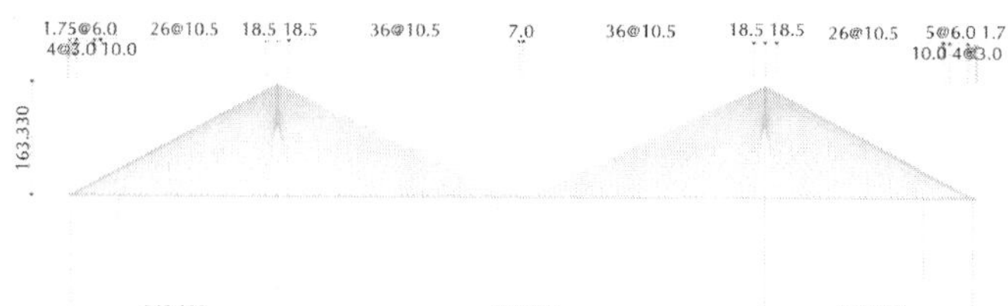

Figure 2: Cable spacing of 10.5 m.

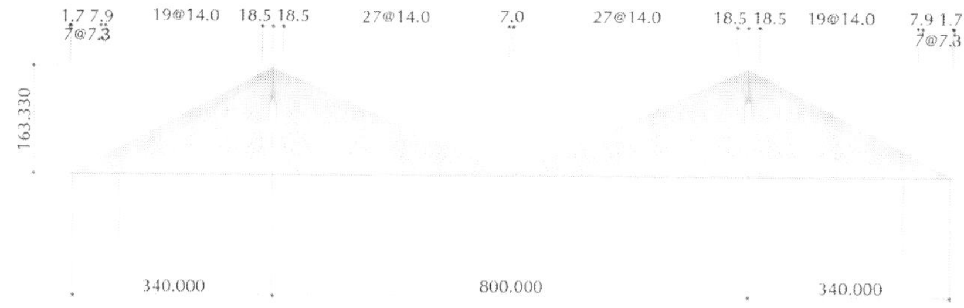

Figure 3: Cable spacing of 14.0 m.

The amount of cables used in each spacing is presented in Figure 4, where the X axis represents the cable spacing. In Figure 4, the values of percentage were calculated relatively on the basis of the amount of the cable at cable spacing 10.5 m. It is known that the total weight of the cables in a cable-stayed bridge does not vary with shorter spacing. In this evaluation, the change is seen to be less than 2%. However, this observation holds only for the amount of cables but does not consider the increase of the number of anchors with the shortening of the cable spacing, which affects the economic efficiency of the bridge.

Tables 1 and 2 arrange the stresses developed in the upper girder under the application of the load combinations involving the permanent and live loads as principal loads, and the replacement and breakage of cable. The compressive stresses occur at the pylon whereas the tensile stresses occur at the center of the main span. Under the combination of the principal loads, the stress developed in the upper girder experiences insignificant change according to the variation of the cable spacing but appears to increase with longer cable spacing under the cable replacement and breakage combinations. The compressive stress at the whole remains below the allowable stress while the tensile stress in the top chord exceeds the allowable stress in all cases indicating the need for strengthening by prestress.

Figure 5 plots the amounts of tendons to be arranged in the cross-section exceeding the allowable tensile stress per cable spacing of the hybrid cable-stayed bridge. These amounts are computed assuming that the tendons installed at mid-span will resist axially by being arranged at the centroid of the cross-section. For the cable spacing of 14.0 m, the prestress stress to be introduced in the upper girder reaches a high value of 53.52 MPa, which represents a surplus in steel wires of

36% compared to the 10.5 m cable spacing. Following, the additional arrangement of steel wires all along the stiffening girder to prepare for extreme events will cause not only the increase of the construction cost but also the extension of the construction period.

Table 1: Compressive stress by allowable stress combination (pylon)

Stress [MPa]		Compressive stress (overdesign factor incl.)		3) Allowable comp. stress	Utilization ratio	
		1) Top chord	Top chord 1)13) 2) Bottom chord	Bottom chord 2)/3)		
Principal loads	7.0 m	-46.01	-55.22	-108.0	0.426	0.511
	10.5 m	-40.72	-54.76	-108.0	0.377	0.507
	14.0 m	-43.13	-71.83	-108.0	0.399	0.665
	7.0 m	-28.58	-30.62	-108.0	0.265	0.284
Cable replacement	10.5 m	-23.91	-30.76	-108.0	0.221	0.285
	14.0 m	-25.16	-44.75	-108.0	0.233	0.414
	7.0 m	-25.99	-28.19	-108.0	0.241	0.261
Cable breakage	10.5 m	-22.39	-27.69	-108.0	0.207	0.256
	14.0 m	-40.61	-61.40	-108.0	0.376	0.569

Table 2: Tensile stress by allowable stress combination (center of main span)

Stress [MPa]		Tensile stress (overdesign factor incl.)		3) Allowable tensile	PS reinforcement stress	
		1) Top chord	2) Bottom chord	stress	Top chord 1)-3)	Bottom chord

Principal loads	7.0 m	10.66	24.76	7.6	3.06	24.76
	10.5 m	11.96	26.51	7.6	4.36	26.51
	14.0 m	15.49	28.33	7.6	7.89	28.33
	7.0 m	8.16	33.98	7.6	0.56	33.98
Cable replacement	10.5 m	11.26	42.65	7.6	3.66	42.65
	14.0 m	15.64	50.10	7.6	8.04	50.10
	7.0 m	5.74	35.56	7.6	0.00	35.56
Cable breakage	10.5 m	8.29	47.33	7.6	0.69	47.33
	14.0 m	13.01	61.52	7.6	5.41	61.52

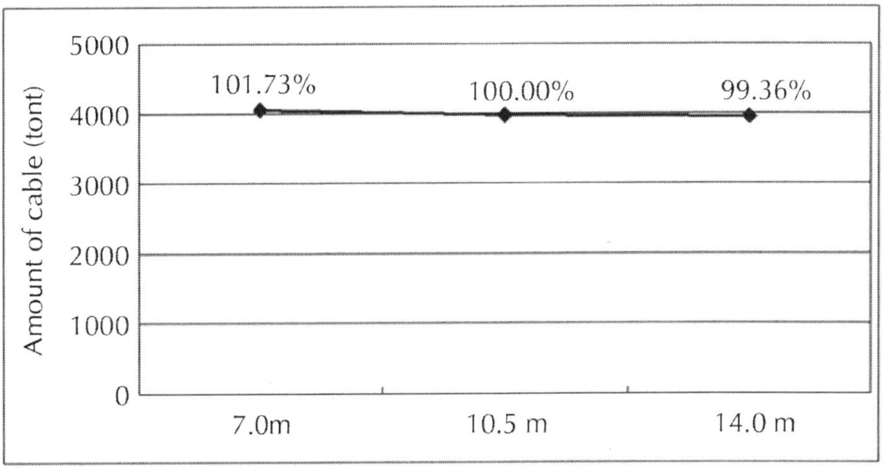

Figure 4: Amount of cables per cable spacing.

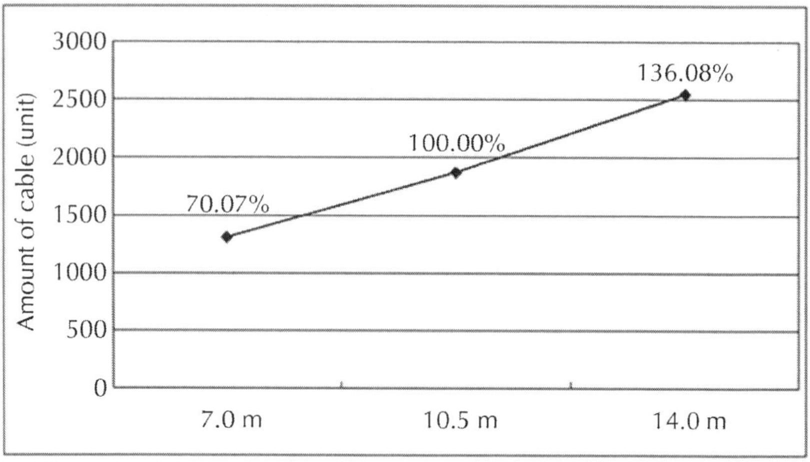

Figure 5: Amount of tendons required per cable spacing.

Figure 6 compares the construction cost according to the cable spacing for the amount of tendons shown in Figure 5 as the sum of construction costs of the upper girder, steel wires, cables (including the anchorages) and pylons. The results indicate that the cable spacing of 10.5 m is the most cost efficient but with a benefit below 1%.

Check of Side Span Length Ratio

Determining the length of the main span is the most important process in the design of the cable-stayed bridge. However, the structural behavior of the whole system varies according to the length of the side spans even if the length of the main span has been decided. This side span length ratio is also influencing significantly the economic efficiency. In this study, the side span length ratio is determined by examining the stresses developed in the girder and the negative reactions occurring in the piers by performing analysis of the whole system subjected to the permanent loads and live loads as the loads governing the design for the cases where the length of the main span is fixed to 800 m and the side span length ratio is 0.4 (320 m), 0.425 (340 m) and 0.5 (400 m).

Figure 7 plots the amount of cable per side span length ratio. It can be seen that the amount of cable increases with larger side span length ratio due to the resulting increase of the length of the whole system. In addition, the amount of cable increases also when the side span length ration decreases because of the increase of the prestress force of the anchor cables needed to control the displacement of the pylons.

The side span length ratio of the cable-stayed bridge has direct effect on the negative reactions occurring at the anchor piers. Even if the small negative stresses do not provoke particular problem in the case of large side span length ratio, counter weight or tie down should be considered in such case. From the examination results, the occurrence of negative reactions at side span length ratio smaller than 0.45 necessitates the consideration of countermeasures (Figure 8).

Tables 3 and 4 arrange the stresses developed in the upper girder under the application of the load combinations involving the principal loads that are the permanent and live loads. It can be observed that the stresses developed in the upper girder subjected to the principal load combination increase with larger side span length ratio and, consequently increase the use of tendons.

Figure 9 plots the amount of tendons necessary per side span length ratio and shows that the required amount of tendons increases with larger side span length ratio.

Table 5 lists the construction cost computed per side span length ratio for the amount of tendons of Figure 9.

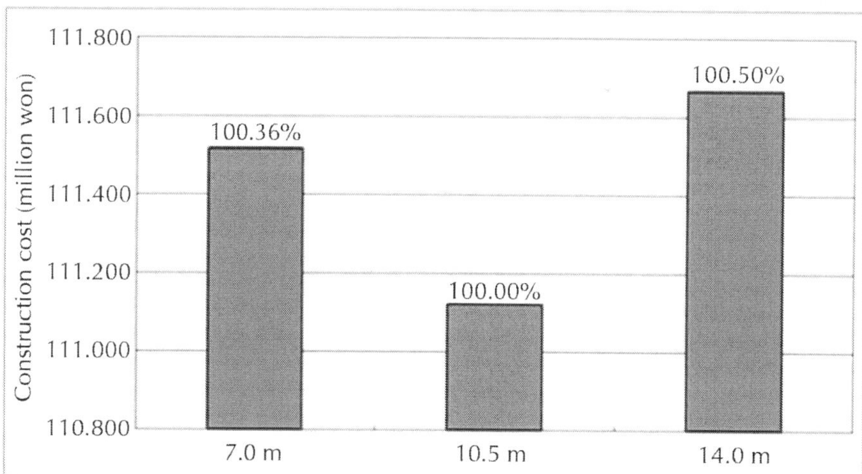

Figure 6: Construction cost per cable spacing.

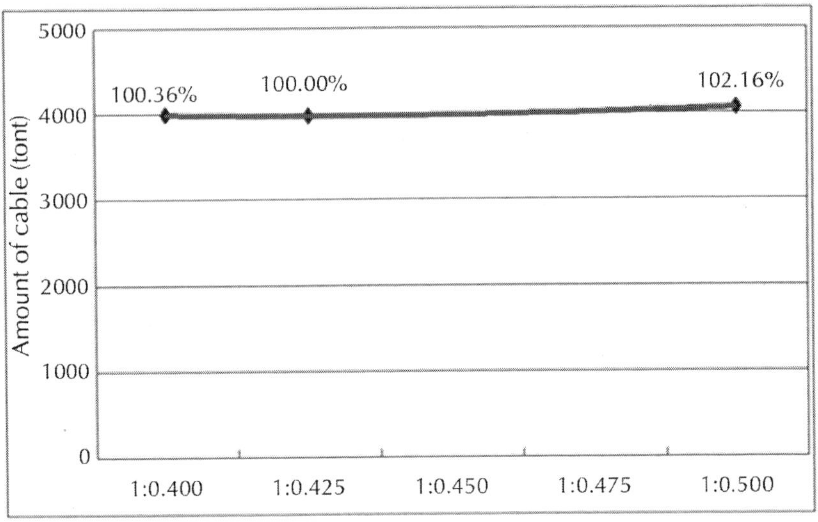

Figure 7: Amount of cable per side span length ratio (ton).

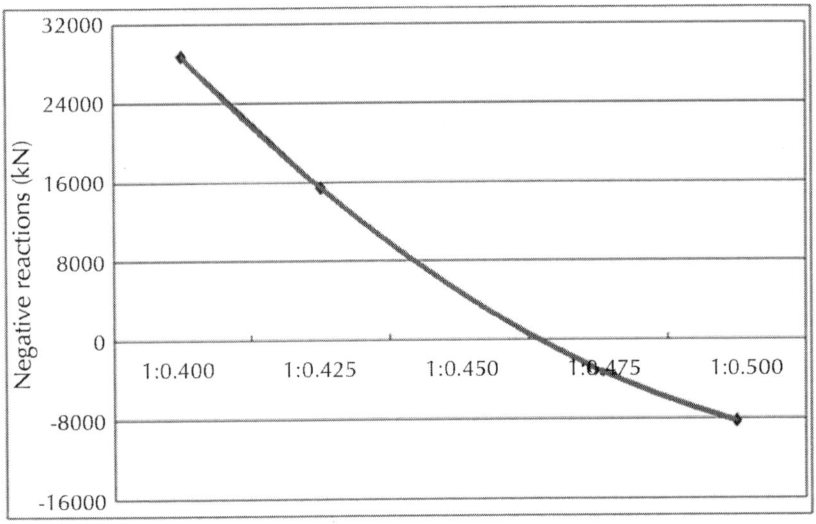

Figure 8: Negative reactions developed per side span length ratio (kN).

Here, the construction cost is the sum of the construction costs of the upper girder, steel wires, cables and pylons. It can be seen that the most economically efficient case corresponds to the side span length ratio of 0.425.

Check with Respect to Pylon's Height

The span-to-rise ratio as the ratio of the ratio of the pylon's height to the length of the main span is an important factor affecting not only the structural efficiency but also the aesthetical appearance of the bridge. In this check, the appropriate span-to-rise ratio is determined by examining the stresses developed in the girder and the sectional forces occurring at the bottom of the pylons by performing analysis of the whole system subjected to the permanent loads and live loads as the loads governing the design for the cases where the length of the main span is fixed to 800 m and varying the height of the pylon.

Table 3: Compressive stress by allowable stress combination (pylon)

Stress [MPa]		Compressive stress		3) Allowable comp. stress	Utilization Ratio	
		1) Top chord	2) Bottom chord		Top chord 1)/3)	Bottom chord 2)/3)
Principal loads	0.400	-40.59	-54.90	-108.0	0.376	0.508
	0.425	-40.72	-54.76	-108.0	0.377	0.507
	0.500	-40.13	-54.55	-108.0	0.372	0.505

Table 4: Tensile stress by allowable stress combination (center of main span)

Stress [MPa]		Tensile stress		3) Allowable tensile stress	PS reinforcement stress	
		1) Top chord	2) Bottom chord		Top chord 1)-3)	Bottom chord
Principal loads	0.400	11.90	25.34			
	0.425	11.96	26.51			
	0.500	13.87	29.00			

Table 5: Quantities of materials and construction costs per side span length ratio

Member	Quantities of materials (tonf)			Construction costs (x10^6 KRW)		
	0.400	0.425	0.500	0.400	0.425	0.500
Cable	3393	3979	4065	42157	40585	41460
Girder	37802	38990	41537	31376	32361	34476

Pylon	53082	53082	53082	31637	31637	31637
Steel wire	630	664	753	6195	6537	7414
Total	-	-	-	111366 (2.315)	111.120 (2.248)	114.987 (2.151)

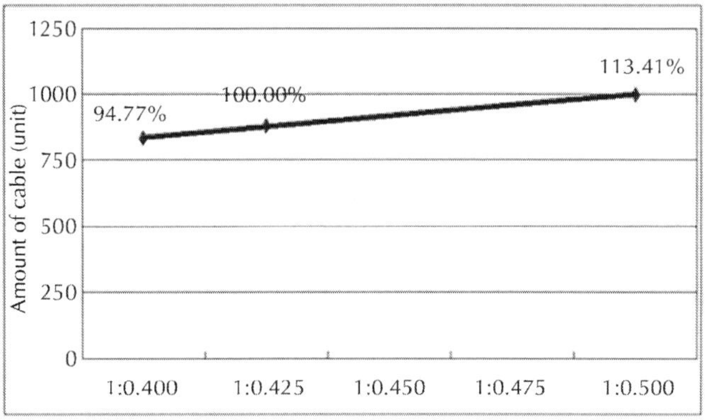

Figure 9: Amount of tendons per side span length ratio.

Figure 10 plots the amount of cables required per span-to-rise ratio. The length of the cables increases with smaller span-to-rise ratio due to the resulting increase of the pylon's height but the efficiency of the cables increases and reduces the amount of cables.

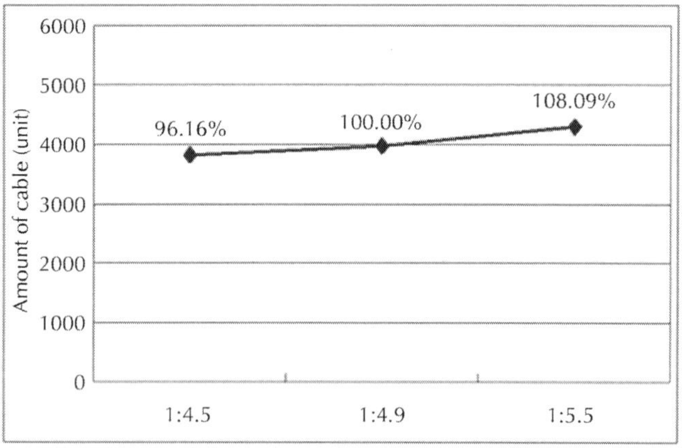

Figure 10: Amount of cables per span-to-rise ratio (ton).

Even if the structural efficiency of the cables appears to be outstanding for higher pylons in the cable-stayed bridge, the quantities of materials required for the pylons increases resulting in longer construction period. Moreover, higher slenderness ratio of the pylon means also vulnerability to buckling and increase of the moment magnification factor at the bottom of the pylon, which finally results in the increase of the cross-section of the pylon. From Figure 11, the buckling safety factor falls down to 1.09 for the span-to-rise ratio of 1:4.5 and indicates that the overall cross-section of the pylon should be increased.

Tables 6 and 7 arrange the stresses developed in the upper girder under the application of the load combinations involving the principal loads that are the permanent and live loads. The stresses in the upper girder subjected to the principal load combination do not exhibit significant difference according to the span-to-rise ratio. Figure 12 plots the amount of tendons necessary per span-to-rise ratio and shows that the span-to-rise ratio of 1:4.5 requires the smallest amount of tendons.

Table 6: Compressive stress by allowable stress combination (pylon)

Stress [MPa]		Compressive stress		3) Allowable comp. stress	Utilization Ratio	
		1) Top chord	1) Top chord		Top chord 1)/3)	Bottom chord 2)/3)
Principal loads	1:4.5	-38.07	-52.21	−108.0	0.353	0.483
	1:4.9	-40.72	-54.76	−108.0	0.377	0.507
	1:5.5	-47.26	-57.91	−108.0	0.438	0.536

Table 7: Tensile stress by allowable stress combination (center of main span)

Stress [MPa]		Tensile stress		3) Allowable tensile stress	PS reinforcement stress	
		1) Top chord	1) Top chord		Top chord 1)-3)	Bottom chord
Principal loads	1:4.5	11.87	26.13	7.6	4.27	26.13
	1:4.9	11.96	26.51	7.6	4.36	26.51
	1:5.5	12.33	26.46	7.6	4.73	26.46

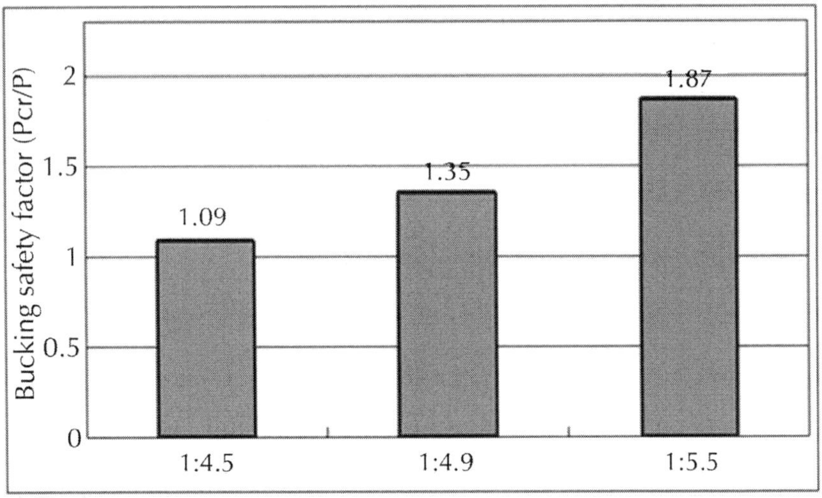

Figure 11: Check of buckling of pylon according to span-torise ratio.

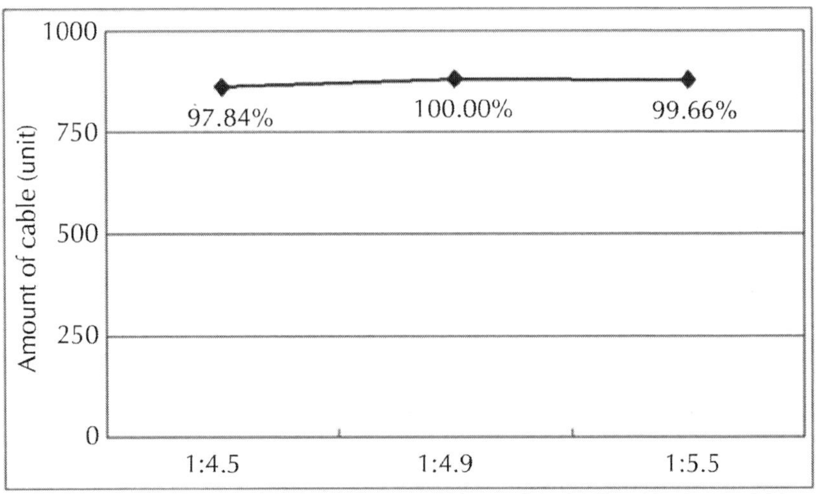

Figure 12: Amount of tendons per span-to-rise ratio.

Table 8 lists the construction cost computed per spanto-rise ratio for the amount of tendons of Figure 12. Here, the construction cost is the sum of the construction costs of the upper girder, steel wires, cables and pylons. It can be seen that the most economically efficient case corresponds to the span-to-rise ratio of 1:4.9.

Table 8: Quantities of materials and construction costs per span-to-rise ratio

Member	Quantities of materials (tonf)			Construction costs		
	1:4.5	1:4.9	1:5.5	1:4.5	1:4.9	1:5.5
Cable	3826	3979	4301	39027	40585	43869
Girder	38,990	38,990	38990	32,361	32361	32361
Pylon	59,630	53,082	51562	35,540	31637	30731
Steel wire	650	664	662	6396	6537	6515
Total	-	-	-	113,324	111,120	113,476

Derivation of Optimal Structural System: Super Bridge 800

Based on these parametric analysis results, a cable spacing of 10.5 m is selected for Super Bridge 800, the hybrid cable-stayed bridge using UHPC girder (Figure 13). The superstructure is an edge girder with depth of 1.735 m and tapered cross beams with height varying from 1.735 m to 3.0 m (Figure 14). One cross-beam with sag of 3.5 m is disposed in each segment. The cables disposed at spacing of 10.5 m are planned to be tensioned after the installation of 3 segments. One segment weighs approximately 1100 kN.

STRUCTURAL CHECK OF SUPER BRIDGE 800

Check Per Construction Stage

The objective of the construction stage analysis is to verify the constructability of the structure and to verify if safety is secured during the erection. Considering the criteria related to the safety of the structure, the appropriateness of the management of the construction equipment and erection sequence as well as the eventual utilization of construction facilities like construction bent and the selection of the position at which they will be used are determined and verified by means of the construction stage analysis.

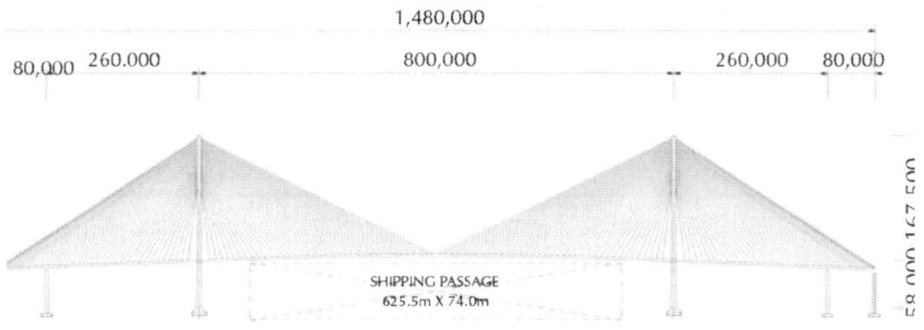

Figure 13: Super Bridge 800.

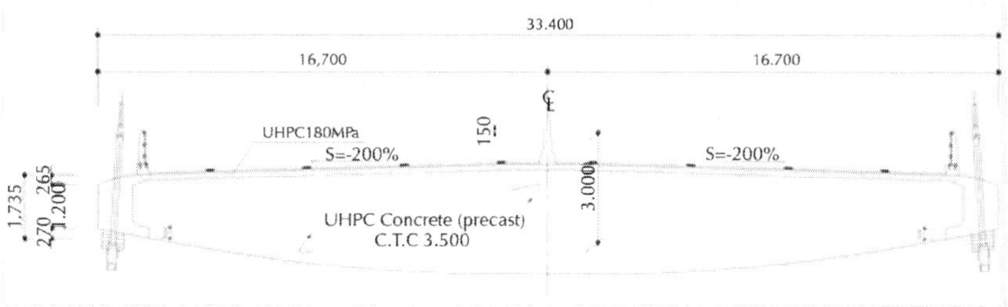

Figure 14: Standard cross-section.

During the erection, the cable-stayed bridge system is featured by the nonlinearities of the cables, stiffening girder and pylons which turn to behave nearly linearly in the completed structure. Therefore, the ranges of the member forces to occur under the load combinations in the completed system are selected at first and considered during the construction stage analysis to verify the tension of the cables at completion. The final tensioning of the cables is obtained through repetitive calculations to put the member forces within the allowable ranges by varying the tension of the cables during the erection.

The analysis was performed by dividing the construction into 280 successive steps. The analysis of the completed system was conducted considering all the member forces at completion and 10,000 days after completion. It was assumed that, when a construction stage was involving several processes, these processes were implemented concurrently.

Figure 15 represents the sectional forces in the girder during the closure of the central span at step 126. Figure 16 shows the sectional forces in the girder during the application of the secondary permanent loads at step 126.

When the length of the cantilever is the longest, known as the most dangerous case during the erection just before the closure of the side spans, safety check during the erection is performed considering all the loads provoking unbalanced moments in the girder. Table 9 arranges the check results relative to the allowable stresses under the application of these unbalanced loads during the erection.

CHECK FOR CABLE REPLACEMENT AND BREAKAGE

The cable replacement is normally checked under the condition that one design lane at proximity of the relevant cable is controlled but all the lanes are loaded considering the vehicle for replacement. The tension of the cables is obtained first with respect to the load combinations. Thereafter, this tension is applied inversely to the structural system to simulate the removed cable [4, 5]. The final tension is finally computed by adding the tension of the cable corresponding to the load combination and the additional tension caused by the replacement. Table 10 lists the stresses developed during the replacement of cable.

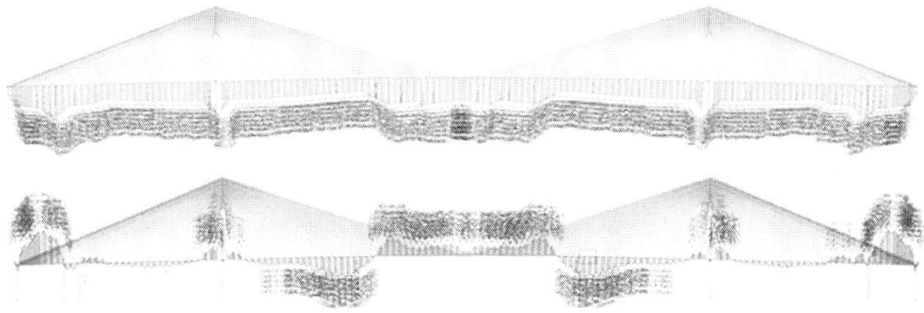

Figure 15: Section forces at step 126 (at closure of central span). (a) Axial forces; (b) Moments.

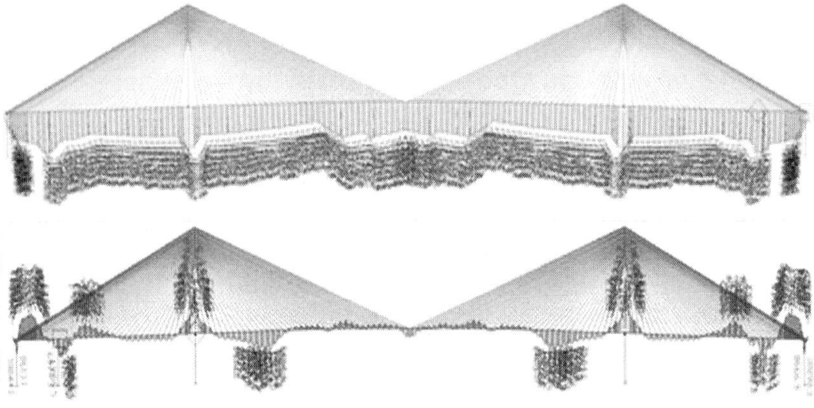

Figure 16: Section forces at step 126 (at application of secondary permanent loads). (a) Axial forces; (b) Moments.

For the cable breakage analysis, the relevant cables are selected at proximity of the sections of the stiffening girder and at several points considered to be vulnerable. The vulnerable sections are positioned at 4 locations that are the cables closest to the pylons, the cables at midspan and 3/4 span and the central span. Since the time needed for cable breakage is in direct correlation with the impact, the check assumes that sudden cable breakage during a period of 0.01 sec to provide conservative results. The damping of the structure is assumed to be 0.02 [6]. Table 11 lists the dynamic analysis results according to the positions of the vulnerable sections.

WIND STABILITY CHECK

The structural safety of the structure is verified during the erection and at completion considering the dynamic wind load effect by means of buffeting analysis for Super Bridge 800. Structural safety is checked by proposing the installation of wind stabilizing cables during the construction stage known to be the most vulnerable to dynamic wind loading.

Table 9: Check of allowable stresses during the application of unbalanced loads

Load case Max.		Stress (MPa)		Allow. stress (MPa)		Stress check		
		Min.	Tension	Comp.	Tension	Comp.	Remark	
Case 1	Top chord	4.782	-21.23	7.6	-108.0	62.9%	19.7%	O.K
	Bottom chord	-2.754	-37.60	0.0	-108.0	-	34.8%	O.K
Load combina-tion	Top chord	0.074	-21.13	7.6	-108.0	0.97%	11.2%	O.K
	Bottom chord	-1.552	-40.32	0.0	-108.0	-	37.3%	O.K

Table 10: Check of allowable stress during cable replacement

Load Max.		Stress (MPa)		Allow. stress (MPa)		Stress check		
		Min.	Tension	Comp.	Tension	Comp.	Remark	
Service load combina-tion	Top chord	-2.031	-48.63	4.0	-108.0	-	45.0%	O.K
	Bottom chord	3.871	-86.73	4.0	-108.0	96.8%	80.3%	O.K

Table 11: Dynamic analysis results at cable breakage

Position	Cable no.	Static displacement (mm)	Max. dynamic displacement (mm)	Dynamic response factor
Near pylon	C12038	4.968	8.130	1.637
1/2 spot	C12056	178.303	295.140	1.655
3/4 spot	C12065	257.770	394.170	1.529
Central span	C12074	208.160	318.814	1.532

The static three-force components and flutter coefficients determined through 2-dimensional wind tunnel test are applied to evaluate the dynamic wind load effect. It is verified that the stiffening girder and pylons of the completed system secure structural stability and serviceability to dynamic wind load. The results of the dynamic wind load check for the main span before closure reveal that the stiffening girder and pylons secure the structural stability for the vertical vibrations as a system where the side spans are closed, and do not need particular vibration control measures. For the structural system before closure of the side spans, it appears that vibration control should be applied for the stiffening girder since the girder in cantilever state during the construction is vulnerable to vertical vibrations caused by the dynamic wind loading. Therefore, wind stabilizing cables are proposed.

Here, the wind cables shall be initially prestressed during the installation to prevent any loss of the tension. The proposed installation of the wind cables plans to cross the cables at 60.5 m with reference to the pylon and dispose the vertical wind cables on the side spans and main span at 12.5 m and 186.5 m with reference to the pylon. The check for this proposal verified that the stiffening girder and pylons secured the structural stability through this solution.

CONCLUSION

This study developed an optimal structural system for the hybrid cable-stayed bridge having its main structural members made of UHPC with compressive strength of 200 MPa and durable lifetime of 200 years. The optimal structural system was derived by conducting analyses related to the cable spacing, side span length ratio and pylon's height, and the derived structural system was verified to secure stability through structural checks. In addition, calculations showed that Super Bridge 800 developed in this study involved amounts of 57,949 tonf and 985 tonf of UHPC and steel wires, respectively. The computation of the costs based on these material quantities and unit prices of 2010 revealed that the construction cost reduced by about 23.3% compared to Incheon Bridge, a cable-stayed bridge of identical size completed in 2009 in Korea and that Super Bridge 800 secures significant cost competitiveness.

ACKNOWLEDGEMENTS

This work was supported by the "Development of accelerated construction technologies for high pylons" project of the Super Long Span Bridge R&D Program. The authors express their gratitude for the support.

REFERENCES

1. Korea Institute of Construction Technology, "Development of Design and Construction System Technology for Hybrid Cable Stayed Bridge," KICT 2009-102, 2009.
2. Korea Institute of Construction Technology, "Development of Design and Construction System Technology for Hybrid Cable Stayed Bridge," KICT 2010-088, 2010.
3. Korea Institute of Construction Technology, "Development of Design and Construction System Technology for Hybrid Cable Stayed Bridge," KICT 2011-076, 2011.
4. Ministry of Construction and Transportation, "Highway Bridge Design Code," 2005.
5. Korean Society of Civil Engineers, "Design Guidelines for Cable-supported Steel Bridges," 2007.
6. C. H. Lee, W. J. Chin, E. S. Choi and Y. J. Kim, "An Experimental Study on the Joints in Ultra High Performance Precast Concrete Segmental Bridges," Journal of the Korea Concrete Institute, Vol. 23, No. 2, 2011, pp. 235-244. doi:10.4334/JKCI.2011.23.2.235

Chapter 10

Earthquake Damage Assessment for RC Structures Based on Fuzzy Sets

Haoxiang He, Maolin Cong, and Yongwei Lv

Beijing Key Lab of Earthquake Engineering and Structural Retrofit, Beijing University of Technology, Beijing 100124, China

ABSTRACT

A global damage index based on multiple linear force-deformation curves in pushover analysis is presented to evaluate the integrated damage of reinforced concrete structure. The modified coefficient is provided considering the cyclic load and hysteresis energy. The number of inelastic cycles and the coefficient of hysteresis energy concentration are also introduced as damage indices. Hence, multiple damage indices about displacement and energy for performance-based design are considered. The relation of multiple damage indices or factors and the fuzzy damage set is presented by comprehensive fuzzy evaluation; hence, a performance-based multiple fuzzy seismic damage-assessment method for reinforced concrete frame structures is established. The method can be accomplished based on pushover analysis, code spectrum, and capacity spectrum method. The fuzzy seismic damage-assessment method is verified through nonlinear analysis four different structures and the corresponding results and assessment conclusions are accurate.

INTRODUCTION

The reinforced concrete structure will inevitably be damaged and enter elastoplasticity deformation stage when it is subjected to strong ground motion. In order to determine the appropriate and economical selective intervention or rehabilitation strategy for an existing structure, it is necessary to assess accurately the actual resistance and the potential failure modes. It is also essential to carry out the damage assessment for a new designed building in order to verify that the performance and capacity meet the target specification of the seismic design guideline. Based on broad categories for advanced seismic design approaches, performance-based seismic design (PBSD) is at the forefront of earthquake engineering research [1–3]. One of the primary aspects of PBSD is the realistic characterization of seismic structural damage and its direct incorporation in the design or performance evaluation methodology. The damage assessment based on PBSD is to ensure that the structures achieve the expected targets for different performance requirements.

Structural ductility, story drift, element and connection rotation, dissipated energy, and fatigue of the structure are the parameters considered for traditional damage assessment [4]. Based on the fundamental parameters, various local and global damage indices have been proposed to evaluate reinforced concrete structures. Research on the global damage indices is essential because the integral damage assessment should be accurate and involve the damage properties of local elements. However, the global damage indices presented until now have the disadvantages such as limited convergence and not considering cumulative damage. In addition, many damage indices are obtained from pushover analysis and capacity spectrum method whose purpose is to evaluate structural performance under earthquake by overprinting seismic response spectrum curve and structural capacity curve [5]. A significant amount of research has shown that capacity spectrum method can simplify computation, consider the nonlinear performance of the structural elements and finally obtain effective results. Because capacity spectrum method has major simplifications, such as a multi-degree of freedom system being transformed into a single-degree of freedom system, nonlinear time-history analysis being replaced by pushover method, and accumulated dissipative effect not sufficiently considered, thus the evaluation is inexact in some cases, especially for irregular or complex structures. Hence, it is necessary to present a comprehensive method to combine multiple damage parameters and obtain more accurate results.

Moreover, the uncertainty always affects the modeling and evaluation of real physical phenomena with different intensity, the damage of structures in earthquake is a vague concept because there exists not explicit classification criterion to evaluate different damaged members and the structure with different damaged information. The values from different damage-assessment methods are also different and even distinct. Hence, the seismic damage assessment and the decision problem are very complex for large and complex structure, which needs the integration based on multiclass information, and the combination of different information is a valid method to reduce the deviation and the uncertainty. In the past decades several alternative mathematical approaches to deal with uncertainty have been developed and the main categories methods include fuzzy set theory, artificial neural networks, interval analysis, and convex modeling.

In classical set theory, the membership of elements in a set is assessed in binary terms according to a bivalent condition; that is, an element either belongs or does not belong to the set. By contrast, fuzzy set theory permits the gradual assessment of the membership of elements in a set [6]. This is described with the aid of a membership function valued in the real unit interval. Fuzzy sets generalize classical sets, since the indicator functions of classical sets are special cases of the membership functions of fuzzy sets, if the latter only takes value

0 or 1. The fuzzy logic and the fuzzy set provide a powerful tool to solve the uncertainty problem and describe the fuzzy characteristics of the target. Especially, fuzzy logic is fit for describing and processing the uncertain information from multiple sources. In recent years, the methods such as the fuzzy logic system, artificial neural networks, and other soft computing methods develop rapidly, which provides a solid theoretical basis for information fusion and system assessment.

In this paper, the multiple damage indices on displacement and energy for reinforced concrete structures are discussed and applied based on performance-based seismic design and the comprehensive evaluation method is proposed by applying fuzzy set theory. The relationship between the multiple damage factors and the fuzzy damage set is established and an effective and accurate seismic damage-assessment method is presented.

DAMAGE-ASSESSMENT INDEX BASED ON DISPLACEMENT AND ENERGY

Damage Index Based on Capacity Curve

The vulnerability of many existing structures may be due to structural weaknesses and low ductility. Common weaknesses in the structural system are due to incomplete load path; strength and stiffness discontinuities, vertical, horizontal and mass irregularities; weak column and strong beam, and eccentricities. Low ductility detailing is characterized as insufficient shear reinforcement, inadequate confinement, and insufficient anchorage length of the beam-reinforcement bars.

The state of damage of a component, a story, or the whole structure may be represented by an index. The damage index is used as an indicator to describe the state of the lateral load-carrying capacity and the reserve capacity of existing structures. Thus, the study on damage index and its availability is necessary.

Some damage indices are calculated for each component of the building (local damage index). The component damage indices may be integrated using a weighting procedure to provide the global damage index for the structure. These damage indices have been formulated using response parameters of the structure that are obtained through analytical evaluation of structural response. There are several techniques and approaches for damage analysis of structures, such as pushover analysis, nonlinear time history analysis, and vulnerability analysis. The typical response-based damage indices include ductility ratio, interstory drift, slope ratio, maximum drift, flexural damage ratio, low cycle fatigue, final softening index and Park-Ang index. The damage indices such as interstory drift and maximum drift, are fundamental and essential for representing the displacement or deformation.

Though various local and global damage indices were presented, the most acceptable damage index is the Park-Ang damage index combining both ductility and cumulative hysteretic energy demand [7, 8]:

$$D_{PA} = \frac{\delta_m}{\delta_u} + \frac{\beta}{\delta_u F_y} \int dE_h \tag{1}$$

Where δ_m is the maximum experienced deformation, δ_u is the ultimate deformation of the element, F_y is the yield strength of the element, $\int dE_h$ is the hysteretic energy absorbed by the element during the response history, and β is the model constant, parameter, which was suggested to be 0.1 for nominal strength deterioration.

The Park-Ang damage index also has deficiencies even though many modified method were proposed [9–12]. A perfect damage index typically normalizes the damage on a scale of 0 to 1, where zero represents undamaged state while unity represents the collapse state of the building. The maximum of the Park-Ang damage index is over 1 and nearly close to 2 in some cases. In addition, the Park-Ang damage index is not directly suitable for nonlinear static pushover analysis or normal capacity spectrum method because the cumulative damage does not occur in this case.

Based on pushover analysis, a stiffness damage index of the global frame was presented by Ghobarah et al. [4]:

$$D_k = 1 - \frac{k_F}{k_I}, \tag{2}$$

where k_I is the initial slope of the base shear-top deflection relationship resulting from the pushover analysis of the frame before subjecting it to the earthquake ground motion and k_F is the initial slope of the same relationship but after subjecting the frame to the earthquake time history. The values of these damage indices range from zero to one depending on the amount of damage experienced.

Although the stiffness damage index has some advantages such as being concise and using pushover analysis twice without any other dynamic analysis, the limitation is also obvious. The damage evaluation is not accurate for moderate damage and collapse stage and the value may exceed 1 and the index is not obtained taking into account cumulative effect. In general, a new and comprehensive damage index needs to be proposed by combining the stiffness damage index and Park-Ang damage index.

Referring to (2), a similar stiffness damage index is proposed as

$$D_c = 1 - \frac{k_c}{k_0}, \tag{3}$$

where k_0 is the initial stiffness of elastic stage and k_c is the current stiffness.

It is worth noticing that the damage index in (2) and (3) can also be transferred as the index presented by Dipasquale and Cakmak [5].

$$D_c = 1 - \frac{T_0^2}{T_c^2}, \tag{4}$$

where T_0^2 is the initial fundamental period and T_c^2 is the current fundamental period. The equation above is suitable especially for damage assessment based on time-history analysis because the instantaneous fundamental period can be acquired conveniently.

According to (3), the relationship of current force F_c, the current displacement u_c and the damage index are as follows

$$F_c = k_0(1-D_c)u_c. \quad (5)$$

Based on the form of (5) the typical skeleton curve or capacity curve of reinforced concrete structure behaves nonlinear or multiple linear, as shown Figure 1, and the current force between the ith point and the i+1th point can be expressed as

$$F_c = k_0u1 + k1(u2-u1) + k2(u3-u2) + \ldots + k_{n-1}(u_n - u_{n-1}) + k_n(u_c - u_n), \quad (6)$$

Where $k_i = (F_{i+1} - F_i)/(u_{i+1} - u_i)$.

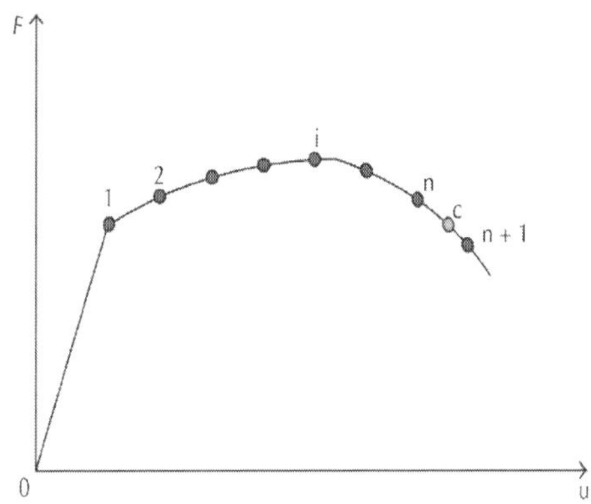

Figure 1: Multiple linear force-deformation curve.

A general global damage index for capacity curve or skeleton curve is obtained by substituting to (6) in (5) with the following result

$$D_c = 1 - \frac{\sum_{i=0}^{n-1} k_i(u_{i+1} - u) + k_n(u_c - u_n)}{k_0u_c}. \quad (7)$$

Compared to (2), the general global damage index presented in this paper has some advantages as follows: (a) the damage index varies from 0 to 1, rigorously, and has definite mechanical significance. (b) The damage index is calculated based on only once capacity curve that is, only one pushover analysis is carried out (c) the damage index is suitable for the capacity curve and the damage index of any displacement or any force can be obtained.

Damage-Index Adjustment Considering Hysteretic Energy

During strong earthquakes, the deformation capacity of structures is reduced due to the cumulative damage caused by cyclic load. However, as the deficiency of pushover analysis, the cumulative damage is not embodied sufficiently and the corresponding modified coefficient is essential. A research on equivalent (reduced) ductility factors has been proposed by Fajfar and the cumulative damage effect is taken into account [13].

The total structure can be simplified as an equivalent single-degree-of-freedom ideal elastoplastic system and the following relations are found or defined: $F_y = mw^2\delta_y, \mu_u = \delta_u / \delta_{y'}$ and $\mu_m = \delta_m / \delta_{y'}$, where F_y is the yield strength of the system, m is the mass, w is the natural frequency, δ_y is the yield displacement, and μ_u is the normal ductility. δ_m Is the monotonic displacement or current displacement and μ_m is the monotonic ductility factor.

The parameter γ, which is a function of dissipated hysteretic energy, maximum displacement, and the natural frequency of the structural system, is introduced as follows:

$$\gamma = \frac{\sqrt{E_H / m}}{w\delta_m} = \frac{1}{\mu_u}\sqrt{\frac{E_H}{m}}, \tag{8}$$

where E_h is the total hysteretic energy. The transformation can be obtained as

$$\frac{E_H}{F_y \delta_y} = \gamma^2 \mu_m^2. \tag{9}$$

Substituting to (8) in (1), the Park-Ang damage index is rewritten as

$$D_{PA} = (1 + \beta \gamma^2 \mu_u) \frac{\mu_u}{\mu_m}. \tag{10}$$

Thus, the relationship between displacement ductility factor on cumulative damage in earthquake and the displacement ductility factor in monotonic load is as follows:

$$\frac{\mu_u}{\mu_m} = \frac{D_{PA}}{(1 + \beta \gamma^2 \mu_u)}. \tag{11}$$

For normal structure, the ratio $\frac{\mu_u}{\mu_m}$ ranges from 0.65 to 0.85. The result of (11) can also be applied to (7) in order to compute the damage index considering dissipated hysteretic energy. For the capacity point which means the crossover point of the capacity curve and the ductility spectrum curve, the corresponding displacement ductility is viewed as μ_u, but the damage status in pushover analysis is related to the monotonic ductility μ_m, so the displacement on the capacity point is advised to be amplified according to (11) in order to represent the actual damage.

Number of Inelastic Cycles

From another point of view, the cyclic demand spectrum for use in preliminary seismic assessment and design is necessary and Kunnath proposed a method based on estimating the number of equivalent cycles at a specified ductility [14]. The methodology is determined using well established relationships between seismic input energy and dissipated hysteretic energy.

If symmetric behavior is assumed, a commonly used idealization of the non-linear behavior is the elastic perfectly-plastic behavior shown in Figure 2. Assuming V_y is the yield force, u_y is the yield displacements and u_{max} is the peak displacements. The energy dissipated by the yielding element per cycle is

$$E_c = 4V_y u_{max}(1-\frac{1}{\mu_c}) = 4\alpha_h V_y u_{max}, \tag{12}$$

Where $\mu_c = u_{max}/u_y$, is the cyclic displacement ductility factor and α_h is an energy shape factor that depends on both the ductility and the shape of the hysteresis loops. For a non-degrading system subjected to N_f full cycles at the same ductility, the total energy dissipated is the product of N_f and E_c, and the following expression between the number of inelastic cycles and dissipated hysteretic energy can be derived:

$$N_f = \frac{E_h}{(4\alpha_h \mu_c V_y u_y)}. \tag{13}$$

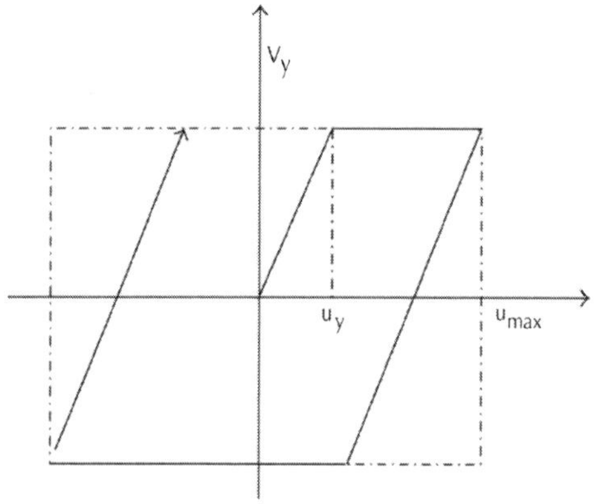

Figure 2: Idealized force-deformation behavior.

Considering the dynamic characteristics and seismic design parameter, the number of inelastic cycles can also be rewritten as

$$N_f = \left(\frac{r_{hl}E_I}{m}\right)\frac{\pi^2}{\alpha_h\mu_c}\left(\frac{R_\mu}{TS_a}\right)^2 = \left(\frac{r_{hl}v_e^2}{2}\right)\frac{\pi^2}{\alpha_h\mu_c}\left(\frac{R\mu}{TS_a}\right)^2, \tag{14}$$

where $r_{hl} = 1.13(\mu_c - 1)^{0.82} / \mu_c$ is the ratio of hysteresis energy to the total input energy E_I, R_μ is the force-reduction factor and the following expression is used

$$R_\mu = \begin{cases} \dfrac{(\mu_c - 1)T}{T_0} + 1, & T \le T_0, \\ \mu_c, & T > T_0, \end{cases}$$
$$T_0 = 0.65T_g(\mu_c)^{0.3}, \tag{15}$$

where T is fundamental period and T_g is the characteristic period.

S_a is the design spectral acceleration; υ_e is the equivalent input energy velocity, which is equal to the Fourier amplitude spectrum of the ground acceleration, and the typical expression is

$$\upsilon_e = \begin{cases} \upsilon_e^*\left(\dfrac{2T}{T_g} - \left[\dfrac{T}{T_g}\right]^2\right), & T \le T_g, \\ \upsilon_e^*\left[\dfrac{T}{T_g}\right]^{-\lambda}, & T > T_g, \end{cases}$$
$$\upsilon_e^* = \frac{\ddot{x}_{g\max}}{4\dot{x}_{g\max}}\sqrt{\frac{\lambda + 0.5}{2\lambda + 2}t_d T_g} = 0.22\pi\sqrt{\frac{t_d}{T_g}}, \tag{16}$$

where υ_e^* is the peak amplification factor for the input energy spectrum, t_d is the duration for the strong motion phase of the ground motion, and λ is a parameter that characterizes the spectral shape of the input energy spectrum for periods longer than the predominant period of the ground motion.

It is demonstrated that the force-deformation characteristics of the system, the ductility-based force-reduction factor R_{μ}, and the ground motion characteristics play a significant role in the cyclic demand imposed on a structure during severe earthquakes.

Coefficient of Hysteresis Energy Concentration

The new and effective global damage indices are described as above. Furthermore, the deformation and hysteresis energy of the special weak stories is important because the structural weakest-part would suffer from severe damages due to its inability to withstand the different types of loads (lateral, vertical and moments) produced by the ground motion [15, 16]. The failure of the weak story can reduce the structural performance though other floors are merely in moderate damage stage. For irregular structures and complex structures, the damage assessment on weak floors is necessary besides the normal analysis carried out.

The coefficient of story yield shear is defined as $\xi_y = V_y / V_{e'}$, where V_y is the yield shear force of a story in the structure, and V_e is the elastic shear force of the story, that is, the shear force for the same deformation if the structure is elastic. The coefficients of story yield shear in each story are equivalent for a regular structure; otherwise the values diverge. The story belongs to a weak story if the coefficient of story yield shear exceeds the threshold value.

The input energy of a multidegree-of-freedom system sustained to an earthquake depends on the structural mass and initial stiffness, in spite of the yield strength value. The story yield first almost is the story that the elastic displacement is the largest and the inelastic interstory drift is also greater than others. The hysteresis energy of the weak story is also greater than that of other stories.

Based on a large number of nonlinear time-history analyses considering various ground motions, fundamental periods, coefficients of story yield shear, and different number of weak stories, the fitting equation of the coefficient of story yield shear and the coefficient of hysteresis energy concentration is as follows:

$$\eta_n = -a_n \xi_y^2 + b_n \xi_y + c_n, \tag{17}$$

where n_n is the story number of structure and an, b_n, and c_n are the parameters varying with n. The statistical values for the structures with different stories are given in Table 1.

Table 1: Statistical values for story yield shear and hysteresis energy concentration

Structure	Parameter		
	a_n	b_n	c_n
5-story	-8.50	6.50	3.00
12-story	-10.0	7.02	4.00
20-story	-12.0	8.06	5.00

The coefficient of story yield shear in each story used to be over 0.3 even if the normal structures are subjected to severe earthquake. Thus the coefficient of hysteresis energy concentration generally decrease when the coefficient of story yield shear increases. Hence, the energy distribution can be obtained through calculating the coefficient of hysteresis energy and measuring the damage-degree by energy is achieved.

Damage Grade and Damage Target

According to the discussion above and the damage indices presented, the global damage index and corresponding adjustment are viewed as an accurate damage-evaluation approach and the number of inelastic cycles can fully embody the integral hysteresis. These two damage indices represent the integral damage states in a reinforced concrete structure. On the other hand, the maximum interstory drift and the coefficient of hysteresis energy concentration can represent the local damage details. Hence, all the four damage indices are advised to be combined and used in damage assessment generally.

Taking into account the damage degree description in Table 2 and the empirical values from nonlinear analysis, the damage degree and the limit values for different damage indices are listed in Table 3.

Table 2: Damage-degree description

Damage degree	Global appearance	Column	**Beam**
Slight	Sporadic occurrence of cracking	Very fine cracks (less than 0.1 mm)	Very fine cracks (less than 0.1 mm)
Minor	Minor cracks; partial crushing of concrete in columns	Visible cracks (0.1 to 0.2 mm)	Visible shear cracks (near support) or tension cracks (at bottom) (0.1 to 0.2 mm)
Moderate	Extensive large cracks; spalling of concrete in weaker elements	Major portion of outer layer of concrete is spalled but core is intact except for hairline cracks (0.2 to 0.5 mm)	Major portion of outer layer of concrete is spalled but core is intact except for hairline cracks (0.2 to 0.5 mm)
Severe	Extensive crashing of concrete; disclosure of buckled reinforcement	Diagonal/torsional cracks in concrete core (0.5 to 3 mm), opening of tie bars, and bucking of longitudinal bars	Reinforcement and concrete bond is broken, there are cracks in the core concrete (0.5 to 3 mm), and shear tie bar have failed
Collapse	Partial or total collapse of building	Crushing of core concrete at joints and relative movement with respect to slab and other columns (cracks > 3 mm)	Crushing of concrete at supports and excessive deflection

Table 3: Damage degree and the limit value for different damage indices

Seismic fortification level	**Frequent earthquake**	**Moderate earthquake**		**Rare earthquake**	
Damage degree	Intact	Minor damage	Moderate damage	Severe damage	Collapse
Global damage index	0~0.10	0.10~0.30	0.30~0.65	0.65~0.85	0.85~1.0
Number of inelastic cycles	—	—	10~20	4~10	0~4
Maximum interstory drift	0~1/450	1/450~1/300	1/300~1/150	1/150~1/50	>1/50
Coefficient of hysteresis energy concentration	1.0~2.073 + 0.046n	2.073 + 0.046n~2.854 + 0.084n	2.854 + 0.084n~3.346 + 0.113n	3.346 + 0.113n~3.554 + 0.132n	>3.554 + 0.132n

DAMAGE-ASSESSMENT BASED ON FUZZY SET

Damage Assessment Model Based on Fuzzy Set

In many practical cases, the structural damage is a fuzzy and vague state because there are no obvious and clear grading marks in the actual status; furthermore, the results of damage assessment based on different damage indices disagree in special cases [17]. These problems may occur when data in Table 3 is used without identification and consideration. Thus, this is more scientific and objective for structural damage according to fuzzy set theory by introducing the concept of fuzzy subset and fuzzy membership [18–21]. In classical set theory, the membership of elements in a set is assessed in binary terms according to a bivalent condition; that is,an element either belongs or does not belong to the set. By contrast, fuzzy set theory permits the gradual assessment of the membership of elements in a set; this is described with the aid of a membership function valued in the real unit interval [0,1]. Fuzzy sets generalize classical sets, since the indicator functions of classical sets are special cases of the membership functions of fuzzy sets, if the latter only take the value 0 or 1. The fuzzy decision fusion method realizes the integration fusion by using fuzzy transform principle and maximum membership principle and considering the various factors relevant to the evaluation objects. In the evaluation process, the factors constitute the factor set as $U = \{U1, U2, \ldots\ldots\ldots, U_m\}$ according to different extent and the decision is expressed as the decision set as $V = \{V1, V2, \ldots\ldots V_n\}$.

First, the single factor decision is made for the single factor $u_i(i = 1, 2, \ldots\ldots, m)$ in the factor set U, then the membership degree r_{ij} in the decision degree $\upsilon_j(j = 1, 2, \ldots\ldots, n)$ is determined according to u_i. Thus, the single factor decision set for the factor is obtained as $r_i = \{r_{i1}, r_{i2}, \ldots\ldots, r_{in}\}$, which is the fuzzy subset of the decision set. Hence, considering all the m factors, the general fuzzy decision matrix is constructed as

$$R=\begin{Bmatrix} r_1 \\ \vdots \\ r_m \end{Bmatrix}=\begin{bmatrix} r_{11} & \cdots & r_{1n} \\ \vdots & \ddots & \vdots \\ r_{m1} & \cdots & r_{mn} \end{bmatrix}, \tag{18}$$

where R is the fuzzy relationship between factor domain and decision domain and r_{ij} is the membership degree. Multi-factor decision should consider the weighted grade for each factor in the fusion process. The fuzzy subset $A=\{a1,a2,...,a_m\}$ is the weighted set in factor domain U, where $a_i(a \le a_i \le 1)$ is the weight value of the single factor u_i in the total decision. When A and R are determined, the fusion is carried out by fuzzy transformation $B = AoR$, B is equivalent fuzzy subset in the evaluation set V, the element $b_j(j=1,2,.....,n)$ is the fusion result for grade , υ_j and o is the fuzzy synthesis operator.

The membership function is the normal distribution function $\mu(x)=e^{-((x-k)/t)^2}$, where the parameters k and t could be determined according to the following principles: (1) the membership value is 1 when the factor is equal to the mean value, (2) the membership value of the top and bottom limitations in the universe of a fuzzy set can be adjusted to 1, and the membership value on the boundary in the universe of a fuzzy set is 0.5.

Damage Assessment Based on Fuzzy Set

Considering various damage indices for reinforced concrete structures, the damage assessment should use one or multilevel fuzzy fusion. According to the importance and the accuracy of each index, the factor fuzzy vector can be determined by the expert experience or analytic hierarchy process, and the membership function is Gaussian model.

Taking a reinforced concrete structure as an example, the global damage index, the number of inelastic cycles, the maximum interstory drift, and the coefficient of hysteresis energy concentration are the main parameters that accurately reflect the structural damage information and they are easy to obtain. Therefore, these parameters are selected in for fuzzy decision.

The weighted values for the parameters can be determined according to the experts grading method or the analytic hierarchy process (AHP). The AHP is a theory of measurement through pairwise comparisons and relies on the judgments of experts to derive priority scales [22]. Considering the importance and calculation accuracy of the grade index about damage state in the factor domain and the result from AHP, the weight value matrix of the damage index is A = [0.4, 0.25, 0.2, 0.15]. The structural damage grade is divided into 5 grades: intact, minor damage, moderate damage, severe damage, and collapse. The damage assessment and grade index are shown in Table 1.

After the fuzzy damage subset is determined, the membership value for the specific damage status in each damage grade can be calculated. In order to meet the requirements of performance-based design, the quantitative description of damage degree is needed. To highlight the dominance hierarchy, the general fuzzy damage index for decision-making is suggested as

$$GFD = \frac{\sum_{i=1}^{4} b_i^k Da_i}{\sum_{i=1}^{4} b_i^k}, \tag{19}$$

where b_i is the membership value of the i-level damage, D_{ai} is the median value in the i-level damage and K is the comprehensive adjustment index, where it is recommended that k is taken as 2.

Damage-Assessment Procedure

In summary, the procedure of damage assessment based on fuzzy set is as follows:

1. Determine the decision set as $V = \{V_1, V_2, \ldots., V_n\}$ in order to describe the different damage grade and the get the equivalent fuzzy subset B in the evaluation set V.
2. Select the factor domain as $U = \{U_1, U_2, \ldots.., U_m\}$; then the damage factors are determined. The corresponding fuzzy subset $A = \{a_1, a_2, \ldots.., a_m\}$ is also established.
3. The nonlinear response is obtained by pushover analysis and capacity spectrum method and the damage indices are calculated.

4. The membership function is calculated and the fuzzy relationship matrix is obtained.
5. The equivalent fuzzy subset in the evaluation set V is calculated by B = AOR.
6. Compute the general fuzzy damage index (GFD) and determine the damage degree of the structure.

CASE STUDIES

In order to verify the damage-assessment method based on fuzzy set, four three-dimensional reinforced concrete frames sustained to strong earthquake are selected to be analyzed by pushover analysis and nonlinear dynamic time history analysis using the three-dimensional nonlinear program CANNY [23]. These 4 structures are as follows: 5-story regular frame, 15-story regular frame, 9-story eccentric frame and 4-story partially symmetric frame, as shown in Figure 3. The concrete compressive strength of the columns is $4 \times 10^7 KN/m^2$, the compressive strength of the beams is $3 \times 10^7 KN/m^2$, the yield strength of the main steel bars is $2.35 \times 10^8 KN/m^2$, the reinforcement ratio of the columns and the beams are 2.2% and 1.8%, respectively. According to the corresponding design demand, the dead loads and live loads in all the structures are $1.0 KN/m^2$ and $2.0 KN/m^2$ respectively, and the loads are transformed as node force in the finite element models. The seismic fortification intensity is 8 and the field type is medium soil.

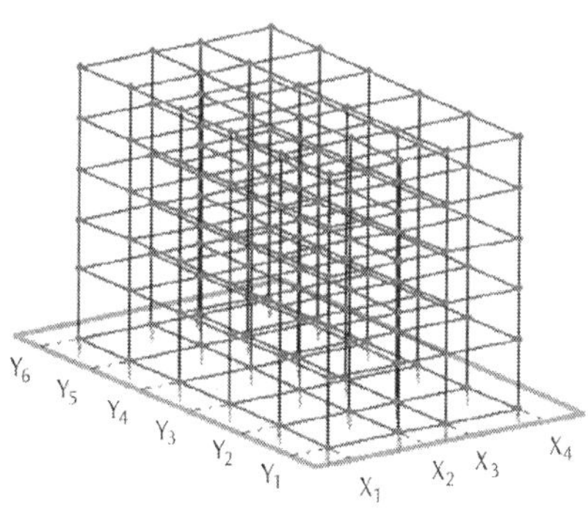

Figure 3: Side elevation of the reinforced concrete frames.

The lateral load distribution of pushover analysis is adopted as the forms based on the stiffness distribution of each story. The capacity curves of the frames are obtained based on the capacity spectrum method. For the capacity spectrum method, an equivalent single-degree-of-freedom (ESDOF) system is a simplistic representation of the actual MDOF model, based on properties of the real structure; a bilinear idealization of the pushover curve can be obtained based on the principle of equality of energy.

For the time-history analysis, each group of waves has two orthogonal horizontal components and one vertical component, the three peak accelerations of each group are adjusted into 4.0 m/s^2, 3.4 m/s^2 and 2.6 m/s^2, respectively, and the duration is 15 seconds.

Taking Frame 1 and Frame 3 as an example, the capacity curves and demand curves are shown in Figures 4and 5 and the global damage curves are in Figures 6 and 7 shown respectively. In the figures, the global damage index on the capacity point is adjusted according to (11). The performance indices are calculated based on (7), (11), (14), and (17). The results are listed in Table 4. Damage assessment based on fuzzy set is carried out and the membership function is established according to the method previously described and specified index in Table 3. For example, the membership function of the global damage index is shown in Figure 8.

Table 4: The results of performance indexes of example structures

Parameter	Global damage index	Maximum interstory drift	Number of inelastic cycles	Coefficient of hysteresis energy concentration	**Equivalent yield displacement (m)**	**Equivalent ductility factor**	**Equivalent period (s)**
Frame 1	0.783	0.019	10.564	4.201	0.183	2.743	0.411
Frame 2	0.762	0.025	3.338	5.334	0.205	2.126	1.968
Frame 3	0.861	0.043	2.278	4.616	0.229	1. 741	1.191
Frame 4	0.726	0.017	6.073	4.238	0.131	2.774	0.752

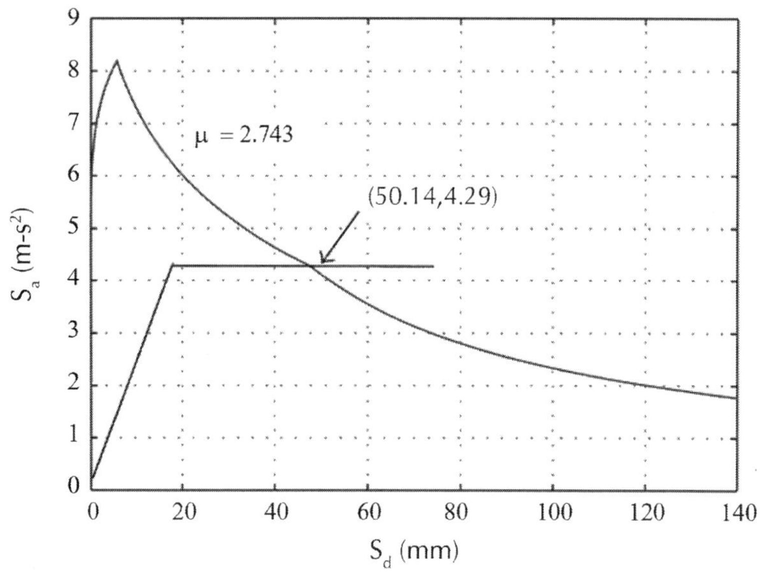

Figure 4: Capacity spectrum curve of frame 1.

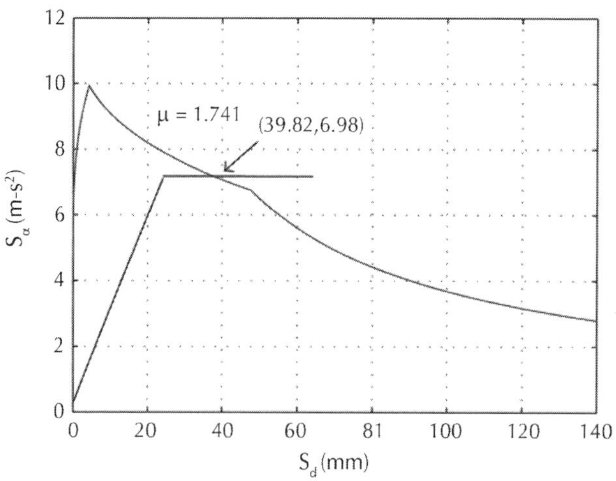

Figure 5: Capacity spectrum curve of frame 3.

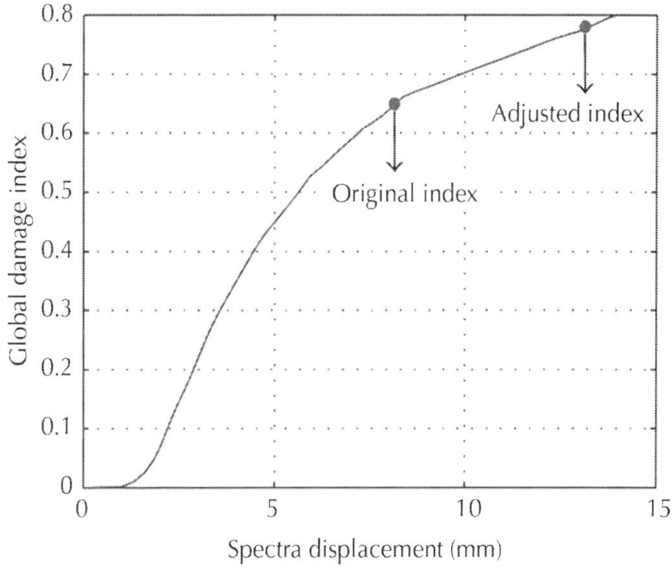

Figure 6: Curves of global damage index of frame 1.

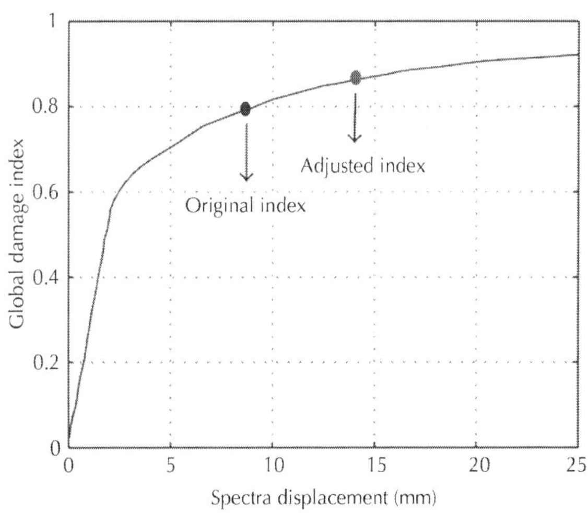

Figure 7: Curves of global damage index of frame 3.

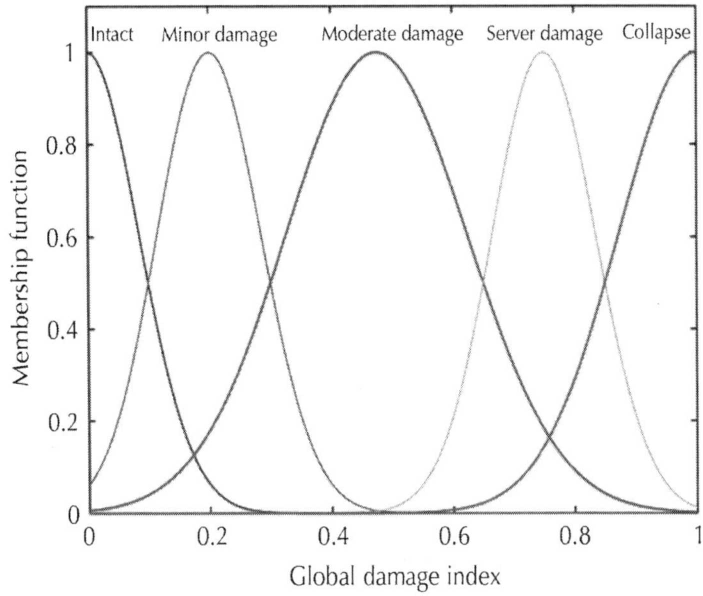

Figure 8: Membership grade of the global damage index.

For frame 1, the fuzzy relation matrix is calculated according to (17) and, combined with the weight value matrix, the fuzzy damage vector is

$$\begin{aligned} B_1 = A \cdot R_1 &= [0.4, 0.25, 0.20, 0.15] \\ &\times \begin{bmatrix} 0 & 0 & 0.116 & 0.926 & 0.236 \\ 0 & 0 & 0.472 & 0.551 & 0.015 \\ 0 & 0 & 0 & 0.618 & 0.447 \\ 0 & 0 & 0.081 & 0.565 & 0.451 \end{bmatrix} \\ &= [0, 0, 0.177, 0.716, 0.255]. \end{aligned} \tag{20}$$

Frame 1 is identified as the structure with severe damage according to the fuzzy damage vector, then collapse and finally moderate damage. The first story is weak story according to the results on interstory drift. Equation (4) is applied as the average damage index for frame 1 in seismic time-history analysis, the value is 0.776, and the damage degree and locations are similar to the results of capacity spectrum analysis. The general fuzzy damage index (GFD) is 0.755 and the final assessment value quantized the damage status accurately.

For frame 3, the fuzzy damage vector is

$$B_3 = A \cdot R_3 = [0.4, 0.25, 0.20, 0.15] \times \begin{bmatrix} 0 & 0 & 0.034 & 0.425 & 0.552 \\ 0 & 0 & 0.011 & 0.172 & 0.761 \\ 0 & 0 & 0 & 0 & 1.000 \\ 0 & 0 & 0 & 0 & 1.000 \end{bmatrix} = [0, 0, 0.016, 0.213, 0.761]. \quad (21)$$

Frame 3 is identified as the structure collapse after earthquake. The top story and the story with irregular stiffness are weak stories according to the results on the interstory drift. The average damage index according to (4) for seismic time-history analysis is 0.856. The general fuzzy damage index (GFD) is 0.858 and the final assessment value is similar to the damage information from the time-history analysis for strong earthquakes.

The important performance results and the fuzzy damage vectors of all the four frames are shown in Table 3and Figure 9. It is evident that the end deformation of the beam and column in frame 2 and frame 3 is obvious, the weak stories are in failure, and the whole performance is deficient and will collapse in strong earthquakes, so the maintenance and retrofitting should be carried out. The damage of frame 1 and frame 4 are both severe but the structures still have the capacity and the performance requirements are met. It is obvious that the seismic damage-assessment method based on fuzzy sets is accurate and effective.

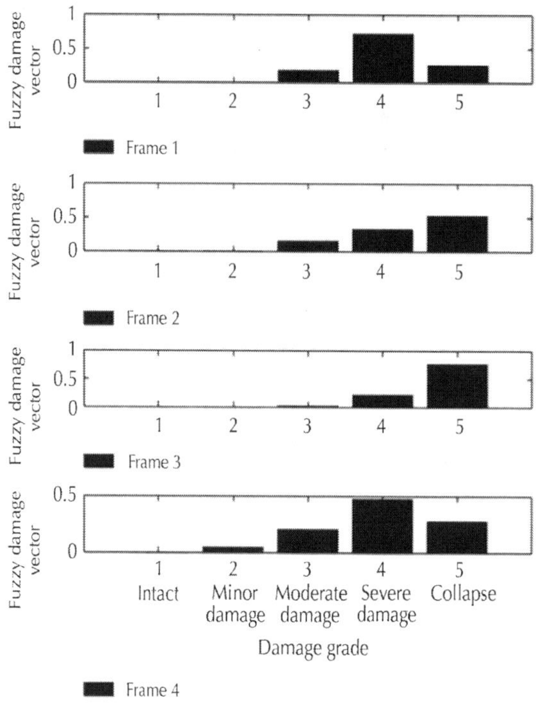

Figure 9: Damage grade for different frames.

CONCLUSION

The damage index is the necessary basis of damage assessment for reinforced concrete structures. Both the global damage index which represents the integrated performance and the local damage index which highlight the damage details are discussed and presented. The response parameters based on displacement and hysteresis energy are all vital especially for the structures subject to earthquake, and the analysis method is suitable for performance-based seismic design and assessment. Though some damage indices such as Park-Ang damage index, are comprehensive and acceptable for normal damage assessment, it is unfortunate that these indices do not converge to unity and are unsuitable for pushover analysis if used directly. A global damage index based on multiple linear force-deformation curves in pushover analysis is presented to evaluate the integrate damage of reinforced concrete structure. The modified coefficient is provided considering the cyclic load and hysteresis energy. The number of inelastic cycles is introduced for the consideration of cyclic loads and hysteresis energy. The coefficient of hysteresis energy concentration is presented as a damage index which reflects the property and failure of the weak story. Thus, the global damage index, the number of inelastic cycles, the maximum interstory drift, and the coefficient of hysteresis energy concentration are all used as the damage indices of damage assessment.

In order to harmonize the different effects and weights, and reduce the uncertainty in the damage assessment, the fuzzy set theory is introduced, and the factor domain, the decision set, the fuzzy subset and the membership function are presented. The relation between the multiple damage indices or factors and the fuzzy damage set is established by comprehensive fuzzy evaluation based on fuzzy mathematics. The fuzzy seismic damage-assessment method is verified through nonlinear analysis for four different structures and the corresponding results and assessment conclusions are accurate, especially when compared with the responses from nonlinear time-history analysis.

The performance-based fuzzy seismic damage-assessment method presented in this paper can be improved and developed in some aspects such as determining the weight values by analytic hierarchy process (AHP) method and considering the multiple dimensional ground motion and the torsional effect in the irregular structures. Furthermore, the performance-based assessment including the foundation and the nonstructural components can also be established based on fuzzy evaluation system.

ACKNOWLEDGMENTS

This work is partially supported by Natural Science Foundation of China under Grant nos. 51108009 and 50878010, Foundation of the Ministry of Education of China for Outstanding Young Teachers in University under Grant no. 20111103120022, and Foundation of Beijing Key Lab of Earthquake Engineering and Structural Retrofit under Grant no. 2013TS02.

REFERENCES

1. SEAOC, Vision 2000: Performance Based Seismic Engineering of Buildings. Sacramento (CA, USA), Structural Engineers Association of California, 1995.
2. A. Ghobarah, "Performance-based design in earthquake engineering: state of development," Engineering Structures, vol. 23, no. 8, pp. 878–884, 2001. View at Publisher · View at Google Scholar · View at Scopus
3. M. J. N. Priestley, G. M. Calvi, and M. J. Kowalsky, Displacement Based Seismic Design of Structures, IUSS Press, Pavia, Italy, 2007.
4. A. Ghobarah, H. Abou-Elfath, and A. Biddah, "Response based damage assessment of structures,"Earthquake Engineering & Structural Dynamics, vol. 28, no. 1, pp. 29–104, 1999.
5. E. Dipasquale and A. S. Cakmak, "On the relation between local and global damage indices," Technical Report NCEER-89-0034, State University of New York at Buffalo, 1989.
6. P. Fajfar, "Capacity spectrum method based on inelastic demand spectra," Earthquake Engineering & Structural Dynamics, vol. 28, no. 5, pp. 979–993, 1999.
7. L. A. Zadeh, "Fuzzy sets," Information and Control, vol. 8, no. 3, pp. 338–353, 1965. View at Scopus
8. Y. J. Park, A. H. S. Ang, and Y. K. Wen, "Seismic damage analysis and damage limiting design of R.C. buildings," Structural Research Serial Report UILU-ENG-84-2007, University of Illinois at Urbana-Champaign, Urbana, Ill, USA, 1984.
9. Y.-J. Park and A. H.-S. Ang, "Mechanistic seismic damage model for reinforced concrete," Journal of Structural Engineering, vol. 111, no. 4, pp. 722–739, 1985. View at Scopus
10. D. Datta and S. Ghosh, "Uniform hazard spectra based on Park-Ang damage index," Journal of Earthquake and Tsunami, vol. 2, no. 3, pp. 241–258, 2008. View at Publisher · View at Google Scholar ·View at Scopus
11. S. Ghosh, D. Datta, and A. A. Katakdhond, "Estimation of the Park-Ang damage index for planar multi-storey frames using equivalent single-degree systems," Engineering Structures, vol.

33, no. 9, pp. 2509–2524, 2011. View at Publisher · View at Google Scholar · View at Scopus

12. A. Massumi and E. Moshtagh, "A new damage index for RC buildings based on variations of nonlinear fundamental period," The Structural Design of Tall and Special Buildings, vol. 22, no. 1, pp. 50–61, 2013.View at Publisher · View at Google Scholar
13. P. Fajfar, "Consistent inelastic design spectra: hysteretic and input energy," Earthquake Engineering & Structural Dynamics, vol. 23, no. 3, pp. 523–537, 1994. View at Scopus
14. S. K. Kunnath and Y. H. Chai, "Cumulative damage-based inelastic cyclic demand spectrum,"Earthquake Engineering & Structural Dynamics, vol. 33, no. 4, pp. 499–520, 2004. View at Publisher ·View at Google Scholar · View at Scopus
15. L. Ye and S. Otani, "Maximum seismic displacement of inelastic systems based on energy concept,"Earthquake Engineering & Structural Dynamics, vol. 28, no. 12, pp. 1483–1499, 1999. View at Scopus
16. C.-C. Chou and C.-M. Uang, "A procedure for evaluating seismic energy demand of framed structures,"Earthquake Engineering & Structural Dynamics, vol. 32, no. 2, pp. 229–244, 2003. View at Publisher ·View at Google Scholar · View at Scopus
17. Y. S. Petryna and W. B. Krätzig, "Compliance-based structural damage measure and its sensitivity to uncertainties," Computers and Structures, vol. 83, no. 14, pp. 1113–1133, 2005. View at Publisher · View at Google Scholar · View at Scopus
18. M. Sugeno, Industrial Applications of Fuzzy Control, Elsevier Science, New York, NY, USA, 1985.
19. E. Cox, The Fuzzy Systems Handbook, Academic Press, New York, NY, USA, 2nd edition, 1999.
20. H. Y. Guo, "Structural damage detection using information fusion technique," Mechanical Systems and Signal Processing, vol. 20, no. 5, pp. 1173–1188, 2006. View at Publisher · View at Google Scholar · View at Scopus
21. I. Andreadis, I. Tsiftzis, and A. Elenas, "Intelligent seismic acceleration signal processing for damage classification in buildings," IEEE Transactions on Instrumentation and

Measurement, vol. 56, no. 5, pp. 1555–1564, 2007. View at Publisher · View at Google Scholar · View at Scopus

22. N. Bhushan and K. Ria, Strategic Decision Making: Applying the Analytic Hierarchy Process, Springer, London, UK, 2004.
23. K. N. Li. CANNY99, Three-Dimensional Nonlinear Dynamic Structural Analysis Computer Program Package, CANNY Consultants PTE, Singapore, 1998.

Citations

CHAPTER 1

Mohamed Elchalakan, Tarek Aly, and Emad Abu-Aisheh, Sustainable concrete with high volume GGBFS to build Masdar City in the UAE, doi:10.1016/j.cscm.2013.11.001.

CHAPTER 2

Y.L. Mo and Rachel Howser Roberts (2013), Carbon Nanofiber Concrete for Damage Detection of Infrastructure, Advances in Nanofibers, Dr. Russel Maguire (Ed.), ISBN: 978-953-51-1209-9, InTech, DOI: 10.5772/57096.

CHAPTER 3

S. O. Nwaubani and A. Katsanos, Effect of Alternative De-icers on the Corrosion Resistance of Reinforced Concrete Bridges and Highway Structures, http://dx.doi.org/10.5772/57551.

CHAPTER 4

Gökçen Firdevs Yucel (2013). Street Furniture and Amenities: Designing the User-Oriented Urban Landscape, Advances in Landscape Architecture, Dr. Murat Ozyavuz (Ed.), ISBN: 978-953-51-1167-2, InTech, DOI: 10.5772/55770.

CHAPTER 5

A.A.A. Molenaar, Durable and Sustainable Road Constructions for Developing Countries, doi:10.1016/j.proeng.2013.03.007.

CHAPTER 6

M. Sigit Darmawan, Ridho Bayuaji, N. A. Husin, and R. B. Anugraha, Case Study of Remaining Service Life Assessment of a Cooling Water Intake Concrete Structure in Indonesia, http://dx.doi.org/10.1155/2014/970393.

CHAPTER 7

Probabilistic failure modelling of reinforced concrete structures subjected to chloride penetration, Caio Gorla Nogueira, Edson Denner Leonel* and Humberto Breves Coda, doi:10.1186/2008-6695-4-10.

CHAPTER 8

V. Patel and N. Shah, A Survey of High Performance Concrete Developments in Civil Engineering Field, Open Journal of Civil Engineering, Vol. 3 No. 2, 2013, pp. 69-79. doi: 10.4236/ojce.2013.32007.

CHAPTER 9

H. Kim, Y. Kim, W. Chin and H. Yoon, "Development of Optimal Structural System for Hybrid Cable-Stayed Bridges Using Ultra High Performance Concrete," Engineering, Vol. 5 No. 9, 2013, pp. 720-728. doi:10.4236/eng.2013.59086.

CHAPTER 10

Haoxiang He, Maolin Cong, and Yongwei Lv, Earthquake Damage Assessment for RC Structures Based on Fuzzy Sets , http://dx.doi.org/10.1155/2013/254865

Index

A

Accelerated damage development 119
Administrations 91
Analytic hierarchy process (AHP) 278, 287
Annual cost 52, 55
Asphalt mixture 122

C

Cable spacing 242, 243, 244, 245, 246, 247, 253, 258
Calcium magnesium acetate (CMA 56
Calcium magnesium acetate (CMA) 56, 81
Calcium nitrite 221, 227
Calcium-silicate-hydrate (CSH) 31
Carbon nanofiber (CNF) 27
Carbon nanotube (CNT) 30
Cementitious material (CM) 6
Chemical admixture 216, 217
Chemical dispersement system 31
Chloride 175, 176, 177, 178, 179, 180, 181, 184, 188, 189, 190, 191, 192, 193, 205, 208, 210, 211, 212, 292
Chloride de-icer 52, 59
Chloride migration coefficient (CMC) 20

Chloride migration coefficient test 21
Civil engineering 214
CNF concrete 32, 36
Compaction effort 123
Compressive strength 137, 139, 140, 141, 143, 144, 146, 148, 149, 162, 164, 166, 170, 173
Concrete core-drilled sample 140, 150
Concrete furniture 90
Concrete member 175
Concrete mix proportion 6
Concrete mixture 221
Concrete structure 176, 179
Concrete structures built 134
Construction and demolition waste (CDW) 124
Construction cost 246, 248, 249, 252, 253
Construction period 244, 251
Construction stage 241, 253, 254, 256
Construction Technology 239, 240, 259
Cooling water intake concrete structure 133, 135
Cooling water intake concrete structure (CWICS) is made of concrete structure supported by steel piles. The concrete structure of CWICS comprises plate (slab), beam, and wall elements. 136
Corrosion measurement 68, 69
Corrosion monitoring 66, 71
Cover depth 178, 181, 188, 189, 190, 191, 192, 193, 194, 195, 196, 198, 199, 200, 201, 202, 205, 207, 208, 209
Creative possibilities 85
Cube axial compressive strength 11, 14, 17
Cubic meter 214

D

De-icing chemicals 59
Demolition waste 124
Demonstrably effective 91
Design compressive 138
Design point 185, 186, 191, 194
Direct current (DC) 28
Drying drum 123
Drying shrinkage test 22

E

Edge loading 120
Electrical resistance variation (ERV) 46
Electrical resistivity 221
Electromagnetic interference (EMI) 34
Elongation flow 31
Environmental Effects 63
Equivalent single-degree-of-freedom (ESDOF) 280
Experimental value 220

F

Fancy equipment 131
Fick's diffusion law 176, 178, 179, 180, 208
Finishing materials 91

First order reliability method (FORM) 178
Fly Ash (FA) 215
Four Probe method 41

G

Geometric design 117, 118, 119, 130
Government sector 53
Gross National Product (GNP) 55
Ground ceramics 218, 220
Ground granulated blast furnace slag (GGBFS) 2, 3, 5

H

Half-cell potential 75, 76
Heavy duty 218
High Performance v, 213, 214, 215, 216, 217, 229, 230, 231, 232, 233, 234, 235, 236, 237, 238, 293
High Performance concrete (HPC) 213
High range water reducer (HRWR) 39
High-range water reducer (HRWR) 31
High Reactivity Met kaolin (HRM) 215
High shear flow 31
Humidity experience 134
Hydraulic Structure 213

I

Important consideration 93

K

Key factor 131

L

LED bulb 91
Linear Polarisation Resistance technique 69
Low-risk de-icer 64

M

Macro crack 29, 30
Maintenance cost 239, 241
Masdar City (MC) 1, 4
Masdar City requirement 4
MDOF model 280
Mechanical model 178, 181, 184, 185, 191, 209
Mechanical strength 10, 14, 17
Molar ratio 62, 63, 79
Multi-wall CNT (MWCNT) 30

N

Nano crack 29
Nano fiber 29, 46, 47
National Association of Corrosion Engineers (NACE) 53

O

OPC replacement 6, 10, 23
Ordinary Portland cement (OPC) 2

P

Pavement life 117, 119, 120
Pavement structure 117, 118, 119, 120, 122, 123, 131
ph shift 73
pH shift measurement 69
Plastic 90
Portland cement 2, 5
Prestressed Concrete Institute (PCI) 32

Pullar-Strecker 72, 80, 84

R

RA aggregate 129
Rapid chloride penetration (RCP) 5
RCP test result 12, 15, 18, 21
Reclaimed asphalt (RA) 128
Recycling construction 124
Reinforced concrete bridge v, 292
Reinforced concrete bridge corrosion 54
Resilient modulus testing 127

S

Scanning electron microscope imaging 70
Scanning electron microscope (SEM) 32
SCCNFC column load path 43
SCC technology 32
SCRC column load path 42
SCSFC column load path 42
Self-consolidating carbon Nano fiber concrete (SCCNFC) 46
Self-consolidating concrete (SCC) 31
Self-consolidating reinforced concrete (SCRC) 36
Self-consolidating steel fiber concrete (SCSFC) 36
Silica fume 216, 217, 218, 220, 221, 222, 223, 224, 225, 226, 227, 228
Silica fume (SF) 215
Single-wall CNT (SWCNT) 30
Slump test 13, 16, 19
Sodium acetate 56, 58, 60, 61, 62, 72
Sodium acetate deicer 60
Steel reinforcement 61
Strategic Highway Research Programmed (SHRP) 214
Street furnishings 86
Street furniture 86, 87, 88, 89, 90, 91, 95, 96, 100, 107, 108, 110, 111, 112, 113
Street Furniture v, 86, 87, 89, 112, 113, 292
Structural life 176, 184, 205
Structural risk 199
Structural safety 181, 190, 196, 199, 201, 209
Structural system 239, 241, 242, 255, 258
Sulphate 216, 217
Superstat potentiostat 69
Supplementary cementitious material (SCM) 3
Surface layer 119, 122
Sustainability 91

T

Tensile strength 217, 219, 222, 224, 225, 226, 228
Tensile stress 30
Thick pavement structure 118
Transport systems 51, 52

U

Ultra high performance concrete (UHPC) 239, 240
Ultra sonication 31
Unidensified Silica fume (USF) 221

V

Van Daveer 71
Viscosity modifying agents (VMA) 32
Viscosity modifying agent (VMA) 39
Volcanic ash (VA) 223

W

Waste receptacle 94
Water cement 78, 80
WA test result 12, 15, 18, 21